Successful professional reviews

for civil engineers

by H. Macdonald Steels

 Thomas Telford

Published by Thomas Telford Publishing, Thomas Telford Ltd, 1 Heron Quay, London E14 4JD.
www.thomastelford.com

Distributors for Thomas Telford books are
USA: ASCE Press, 1801 Alexander Bell Drive, Reston, VA 20191-4400
Japan: Maruzen Co. Ltd, Book Department, 3–10 Nihonbashi 2-chome, Chuo-ku, Tokyo 103
Australia: DA Books and Journals, 648 Whitehorse Road, Mitcham 3132, Victoria

First edition published 1997
Second edition published 2006

Also available from Thomas Telford Books
Dynamic mentoring for civil engineers. H. Macdonald Steels. ISBN 0 7277 3003 7
Eective training for civil engineers. H. Macdonald Steels. ISBN 0 7277 2709 5

A catalogue record for this book is available from the British Library

ISBN: 0 7277 3457 1

© Author and Thomas Telford Limited 2006

Typeset by Academic + Technical, Bristol
Printed and bound in Great Britain by MPG Books, Bodmin, Cornwall

'My task which I am trying to achieve is, by the power of the written word to make you hear, to make you feel – it is, before all to make you *see*. That – and no more, and it is everything.'

Joseph Conrad, *The Nigger of the Narcissus*
from *Four Tales,* Oxford University Press, 1954

Acknowledgements

Yet again, I am indebted to so many people:

❏ the Reviewers, who good-heartedly continue to share their experience and thoughts. Some even found the time to comment on drafts;

❏ the Institution, which allowed me privileged access as the *ICE3000 series* worked its way through the committees, and their membership sta, who drew my attention to the many changes on the way;

❏ my colleague Patrick Waterhouse at Bowdon Consulting, with whom so much of the advice and guidance has been developed for courses and seminars over the past few years;

❏ my family, particularly Ann, who continue to humour, encourage and support me, almost without complaint.

Preface to second edition

The advice and guidance in this edition are in accordance with the *ICE3000 series Routes to Membership*, first published in January 2006.

None of the reasons for writing the first edition of this book in 1997 have gone away; a few have certainly receded, but others have been added. That the need is still there suggests that trainees and their mentors continue to have difficulty understanding the purpose and process of a peer review, despite strenuous efforts by the Institution and the previous edition of this book. Myths and misconceptions still abound; this book sweeps them all away and exposes the real target and how to hit it.

The educational bases for the various classes of membership remain the same as first established in 1997, but there are routes available for anyone, starting from anywhere, to gain a class of membership appropriate to their abilities and contribution. The differentiation between the two Grades of engineer reiterated by the Engineering Council in 2004, both encompassed by Membership (MICE), is better defined. There is now a qualifying class of membership for specialist scientists and technologists upon whom our industry relies, but who are not civil engineers.

The advice in this book covers all the routes to professional qualification, including the progressive route which ought to lead to quicker progress for graduates. The guidance is applicable to potential Technician Members, but they must modify the detail to suit their levels of expertise.

Every effort has been made to avoid mistakes; any that remain are my responsibility. Candidates and their mentors are urged to make their own judgements on the regulations and guidance from the Institution current at the time of submission, and not to rely solely on this book.

H. Macdonald Steels
January 2006

About the author

As Regional Training (later Liaison) Officer for the Institution of Civil Engineers (ICE) from 1988 to 2003, Mac guided some 4500 trainees towards professional qualification, not only with ICE but also with kindred Institutions. Before that, he successfully trained many colleagues towards professional qualifications in a broad range of civil engineering, in both public and private sectors.

He now lectures widely in the UK and overseas on the workplace development of excellence in civil engineering and on preparation for the professional reviews. He is a consultant on change management and the motivation and inspiration of engineering staff and an outsourced Supervising Civil Engineer.

In November 2003, Mac was awarded the ICE's Garth Watson Gold Medal for services to the Institution. The citation particularly noted his 'unique contribution to the recognition and maintenance of appropriate professional standards in civil engineering and civil engineers'.

Contents

Chapter 1

The changing role of the profession

It is very easy in the hurly-burly of everyday life to forget the overriding philosophical nature of our work. The monarch, in the Institution's Royal Charter, states emphatically that

> 'many important and public and private works and services in the United Kingdom and overseas which contribute to the well-being of mankind are dependent on civil engineers'.

Our profession is thus acknowledged as being at the very heart of civilised society. You, as a relatively recent entrant to our profession, must occasionally lift your heart and mind above the mundane day-to-day activities which occupy you, to realise what a profound and far-reaching effect every aspect of your work is having. The Reviewers will expect this breadth of understanding beyond your immediate role.

The Engineering Council's generic documents state that both Chartered and Incorporated members must

> 'make a personal commitment to live by the Code for Professional Conduct, recognising obligations to society, the profession and the environment.'

> $(EC^{UK} SPEC2004)$

Whatever we do, from the largest dam to the smallest patch repair, affects people and the environment. We all seem to remember that part of our Royal Charter which originally defined the profession as 'The Art of directing the great forces of power in Nature for the use and convenience of man', but how many of us have read on, to the statement that our job requires

> 'a high degree of knowledge and judgement in the use of scarce resources, care for the environment and in the interests of public health and safety'?

This, to me, is the most succinct summary I know of the true nature of our business. Even the reference to 'the use of scarce resources' anticipates the greater emphasis on sustainable development, now an intrinsic part of the review process.

The Institution's Royal Charter goes on to require it to

> 'ascertain persons who by proper training and experience are qualified to carry out such works',

which it does through the Professional Reviews. These have changed significantly since the first edition of this book was published in 1997, to accommodate a much greater range of candidates than the traditional 'design and build' stereotype.

All these rather grand ideas may seem a million miles away from the day-to-day problems faced by you as a civil engineer. But each one of us must never lose sight of these ideals and must seize every opportunity to express views and opinions based upon them.

Throughout my working life, UK civil engineers have moaned about their lack of status in society. If there is such a thing as status (and I, for one, am not at all sure what it entails) then we must help the public to understand that their well-being, indeed (one could easily argue) the well-being of the entire planet, is dependent on our judgements.

The Institution is intent on protecting these high and responsible ethics and the criteria of its member review system reflect the global responsibilities which this entails.

> **These high ideals must be a motivating force behind your preparation, submission and performance on the day.**

Sustainability

On behalf of its members the Institution is already taking a leading role in the debates on the environment and what has become known as sustainability, 'meeting the needs of the present without compromising the ability of future generations to meet their own needs' (The World Commission on Environment and Development, 1987). Note the use of the word 'needs', not 'wants' or 'desires'. In highly developed parts of the world, it is easy for people to forget what we actually need, as distinct from what we believe we need.

As always, needs are a matter for judgement and persuasion. Civil engineers are in the business of risk management, and so make similar judgements every day. We are therefore ideally experienced to take on this added responsibility. We must all, however, move away from our former perceptions – no longer 'predict and provide' but 'target and manage', and then persuade the public of the validity of our sensible pragmatism.

Sustainability is a global concept both in absolute terms and in the context of sharing resources. All must act in concert, something which thousands of years of history suggest is not possible. But there are encouraging signs that the world is realising that individual elements within it cannot act in isolation. More than ever before, the world is trying to seek compromise, to stop warfare and combat sickness and famine, even starting to conserve resources – granted not very effectively so far but, as yet, there is not a lot of expertise available.

Civil engineers, with their unique set of skills, particularly their ability to take the broad view and seek acceptable compromise, are uniquely equipped to play a major role in the pragmatic interpretation of sustainability. Much of our society has vested interests – what former ICE President David Green described as 'single issue politics' – the uncompromising pressure group. Civil engineers can rise above these vested interests, to utilise scarce resources, care for the environment and protect the safety and health of the public.

Commitment to sustainable development is a key performance indicator in the review process.

Environment

The real concept of a need for the United Kingdom to improve or protect its environment first arose in the early 1800s. The dramatic effects of the Industrial Revolution quickly made concerned people realise that there was a necessity for legislation to limit the physically dangerous and harmful effects of many manufacturing processes and to reduce the risks inherent in a rapidly increasing population migrating into the new towns (the population of the UK in 1800 was 10.5 million, compared with nearly 60 million today).

What happened in the mid-nineteenth century was an enormous surge in the work necessary to build an infrastructure capable of

❏ realising people's expectations, as enlarged by the revolution;

❏ achieving the standards prescribed by the new laws.

These new requirements led to the great sewerage and water supply schemes, improvements in transportation and the consequential development of strong municipal authorities. This presented a great opportunity to the embryonic British civil engineering profession, whose members responded admirably – with vision, creativity, ingenuity and courage. Almost everyone can recite the names of many of these engineers, who still catch the public imagination today.

The current flurry of fundamental change, prompted by the release of market forces in this country, coupled with the technological revolution, is spawning another burst of legislation to rectify the new social and environmental problems it is creating, just as it did after the earlier Industrial Revolution. This time around there is not the same scope for massive infrastructure works; we are reaching saturation point in this relatively small island, as indeed are other countries. But there is perceptible public pressure for improvements and rethinking of the very basis of our existing infrastructure. There is a growing realisation that rapid and unrestricted industrial development is unsustainable in the longer term.

There is an enormous need for highly competent and skilled engineers, not purely to continue to extend, but to make the best possible use of the existing infrastructure. We must improve its efficiency, with relatively minor capital works, and come up with innovative and ingenious solutions to the many problems being created by increased expectations, a burgeoning world population and environmental change.

Meeting the needs of today without compromising the needs of future generations requires a different outlook, altered skills and all the vision, lateral thinking and ingenuity we can muster! Our qualification system has gone through a painful and difficult transition to provide for this growing need.

A successful review requires you to demonstrate that you have the capacity to rise to this environmental challenge.

Detailed changes to the review system will, and must, continue, to ensure that those becoming professionally qualified meet the developing needs of our industry and the demanding society we serve. But the fundamental philosophy behind the reviews (as specified in our Royal Charter) will remain constant and must be kept in mind throughout.

Chapter 2

Wider membership

Broad descriptors

The range of abilities which the public, society and clients now require of professional engineers is very different to those which were necessary only a couple of generations ago. As a profession, more of us are moving back up the decision-making chain, no longer merely carrying out the wishes of others ('on tap but never on top'), but playing an increasing (and arguably long overdue) role in the process of deciding what is appropriate.

In the UK, with a well-developed infrastructure, the emphasis has moved to managing and maintaining the infrastructure, seeking ways to make what we already have work more efficiently and effectively, reducing the need to build new (the 'predict and provide' of 40 years ago). Obviously some new infrastructure will always be needed, even if only to replace that which is worn out or no longer appropriate.

With an increasing emphasis on sustainable development, satisfying the needs of today without compromising the ability of future generations to meet their needs, the UK civil engineer is supremely well placed to make the distinction between the 'needs', and the wants or desires, of an increasingly affluent society – a Western world which has developed well beyond the basic necessities of clean water, food and shelter, and which rising economies, understandably, now seek to emulate. Can we, for example, help these emerging economies to avoid some of the dreadful mistakes our Western society has made?

So the emphasis of the business is moving further towards the maintenance and the efficient and better usage of existing assets; towards revolutionary and sustainable solutions to ever-bigger environmental, social and geographical problems, well beyond what has been considered traditional ('design and build') civil engineering. Challenging statements such as this, from the erstwhile Chief Executive of Atkins in *New Civil Engineer*, 2 September 2004:

> 'We must understand that the worst thing for any client is having to build something.'

emphasise that these global problems require innovation, creativity, resourcefulness and leadership like never before.

The review criteria for Membership (MICE) of the Institution are capable of wide interpretation, while still being applicable to the more traditional engineer. Look at them very carefully; it is clear (but still the rumours persist) that there is no Institution requirement for specific experience in any particular work environment. Nowhere do the criteria refer to 'construction', 'reinforced concrete', 'site' or 'design', 'bills of quantities' or 'rates build-up'. More importantly perhaps, confidence with analysis software does not, of itself, necessarily demonstrate an understanding of technical principles. Nor is it necessary to do analysis in order to develop an understanding of such principles!

I have personally been associated with engineers who have been successful at the professional reviews from employers well beyond civil engineering companies: house builders, estate agents, telephone companies, commercial insurers, a merchant bank and even, in one notable case, a firm of solicitors. These are, without doubt, extreme examples of highly specialised engineering, but they do emphasise the flexibility inherent in the Institution's criteria.

The Institution's current training and review system is extraordinarily transferable across disciplines well beyond traditional civil engineering. The capabilities being developed are applicable in many other fields of work; they are what is known as transferable skills. For years, enlightened firms have been using a similar system to train engineers for many other Institutions. Training schemes embracing civil, chemical, production, mechanical and electrical engineers became almost routine during my time as a Regional Liaison Officer (RLO).

I found some firms which were transferring the system to quantity surveying, even accountants! Also very encouraging was the number of foreign companies who, realising the efficiency and quality of the Institution system, exported it back to their homeland to assist in the development of their indigenous engineers.

It is vital that, as we move further and further away from traditional engineering, everyone concerned, both trainees and their mentors, as well as employers and Reviewers, recognise that old-fashioned stereotypes are no longer the mainstream. Full use must be made of the flexibility which the Institution, with great foresight, built into its training and review systems 15 years ago. Then we really will start to deliver the Institution's stated aim – a wider membership.

Another complementary method is to recognise that there are persons who make significant intellectual and practical contributions to civil engineering, but who do not have the educational base which would enable them to become qualified as professional civil engineers. So the Institution has created a class of membership which has the same rigour, but which replaces engineering principles with technical or scientific principles, while the other criteria remain the same. This is the new Associate Member.

The Associate Member

The huge social and environmental problems mentioned earlier require specialist skills well beyond those needed for design and build: such diverse things as environmental management, statistical probability, ecology and social understanding, as well as a deeper understanding of the scientific bases on which all these rely. So the breadth of expertise must also grow.

The Institution has recognised that civil engineering needs such skills, and that they are not the sole prerogative of civil engineers.

It has therefore developed wider routes to qualification and offers possible membership for these 'new' sciences and technologies, to embrace people who are making a significant contribution to our profession, but who do not necessarily have, and may not want, civil engineering academic qualifications. This book is as applicable to them as it is to mainstream engineers.

Chapter 3

Surmounting
historical obstacles

The continuing dichotomy of IEng v. CEng

The attributes necessary for a civil engineer set out by the Engineering Council (EC) and developed by the ICE reflect the fundamental changes in our role, but our industry has been slow to catch up. Many senior mentors are still training graduates much as they were trained themselves, on the assumption that nothing has changed and that all their graduates will, almost inevitably, become Chartered Members.

The historical mainstream member of our profession is best matched by the EC description of the Incorporated Engineer, a highly competent technical manager. Yet still the majority of the industry continues to resist the change (after some 30 plus years) and continues to presume that anyone who can demonstrate reasonable engineering competence will become Chartered.

As recently as 2004, in $EC^{UK}SPEC2004$, the Engineering Council reiterated that

> 'whilst there have been some difficulties in securing for IEng the same degree of recognition which the CEng title has secured, there is good evidence that in the majority of industries, the two categories are recognised and the differences between them are understood'.

There is therefore no plan, despite continued resistance in civil engineering, to stray from the avowed intention that the majority of professional engineers in future will be IEng, in line with *all* the other countries in the (enlarged) European Union, where the Technical Engineer (our IEng) is generally held in very high regard. The corollary is surely that there will be fewer Chartered Engineers. Both are vital components of any engineering team – different, but each of equal value.

Both Members are vital components of any engineering team, crudely divided into 'doers and managers of the doing' and those who 'decide what needs to be done and persuade others of its appropriateness', although there is inevitably a considerable overlap. This is why both Chartered Engineers and Incorporated Engineers are now designated Corporate Members, MICE (CEng or IEng).

In the beginning, recruitment should reflect the balance between the two kinds of engineer, dependent on existing and anticipated workload. Salary scales and promotion prospects must be developed and publicised to recognise the particular contribution of IEng quality engineers if they are to be retained and recognised as key members of the team.

At the time most current graduates feel ready for review, it is probable that they are somewhere between the two classes. Having left university with a traditional capability in technical analysis, many will have applied this in the growing context of resource availability and environmental, social and economic considerations. It is therefore likely that they will satisfy the requirements of the Member Professional Review within a realistic timescale (of, perhaps, three–four years). Those that then go on to develop the wider abilities in social and economic decision making will develop the attributes expected of a Chartered Member.

It is imperative that a candidate is clear on the requirements and chooses to apply for the class of membership in which they can most readily demonstrate the requisite attributes. Each sponsoring firm or engineer must positively contribute to deciding which is the most appropriate designation for each individual approaching the professional review.

Recruitment

Many employers continue to recruit Honours degree graduates to do a basic, largely technical, job and appear not to think too far beyond that. There is, sometimes, a subconscious presumption that the better ones will eventually rise to the top, to provide succession management at some undetermined time in the future. Such random and inefficient staff management is, unsurprisingly, commensurate with high graduate turnover and lack of enthusiasm and morale, and must be extremely expensive. Few firms attempt to measure the true cost of losing individual technical staff, but some do, and the figures they produce are alarming (direct costs 'in the region of £43 000'; conference, March 2004).

In my experience, few firms recruit for management succession in the future, but merely to satisfy current needs. The job for which many graduates are still being recruited is essentially that of a technician – a skilled user of techniques and technical systems, mostly the manipulation of sophisticated software or routine exercises such as setting-out (albeit carrying significant responsibility) and quality control.

It is still not unusual to meet graduates who have spent five, six, even seven years repetitively punching data into a computer for things like water distribution network or structural analysis. Taking the processed data and using it as information for the development of a design is a prerequisite, if you are going to develop and demonstrate the required capabilities of a professional engineer. If you believe that, because you are an Honours degree graduate, you will somehow inevitably become Chartered whatever your experience, then your expectation is unrealistic, doomed either by your own lack of ambition, or by your employer. The obvious remedy for the latter is to seek opportunities elsewhere!

Employment

Our business is full of people who work hard, put in long hours and 'keep their heads down'. They are usually very focused on their job and do not seek much in the way of further challenges beyond it.

Such persons are vital to the success of any organisation, consistently delivering their departmental objectives and extremely competent in what they do. Their value is inestimable; proper career structures must be in place to ensure that they are rewarded appropriately if they are to be retained in that role.

These conscientious and capable engineers are the backbone of our industry. But their qualities are inconsistent with those set down by the Engineering Council for a Chartered Engineer, which require engineers to be clearly attuned to the fundamental and overriding needs of the society they serve. They comply with the attributes of an Incorporated Engineer, but unfortunately there are few role models in the hierarchy above them, so they see little point in becoming professionally qualified. Only when companies recognise this, develop career progressions, and actively market the possibilities, will the situation change.

If you want to be a member of the relatively small group of Chartered Engineers, then you must be much more aware of the wider aspects of our profession and of the involvement of your employer in the full business of civil engineering. From an early stage in your career development, you will be demonstrating clear signs of wide vision and strong leadership, wanting speedy and successful outcomes, seeking opportunities both for yourself and your employer, aware of the whole range of influences on your work well beyond the technical. You must define and drive your own career aspirations, not wait for your employer to decide.

Prejudices

For many years, the only qualifying class of membership of the Institution of Civil Engineers was as a Chartered Engineer. Only in 1989, did it include the Incorporated Engineer as an additional class of Member, and it was as recently as 1997 that both Chartered and Incorporated Engineers were made Corporate Members.

This widening of the membership did cause some very illustrious engineers to have serious misgivings; indeed some took the drastic step of resigning from the Institution. This entrenched intellectual exclusiveness is still clouding the judgement of many employers, such that any engineer who is reasonably competent, and has a degree, is still perceived as eligible to become a Chartered Engineer. This attitude is taking a long time to change, even within the Institution itself, but changing it is, and all employers need to recognise this by recruiting, developing and qualifying a range of people appropriate for the needs of their business.

Expectations

There is no disgrace whatsoever in graduates who feel comfortable doing so remaining within a technical role, but they cannot expect to become Chartered Engineers. There is sometimes an unrealistic belief that, merely because they have satisfied the basic academic requirement, progression to Chartered status is somehow inevitable.

Nor must their employers encourage the belief that they can, in cases where it is clear that such progression is unlikely 'for the foreseeable future' (no employer should ever categorically rule out late development). Rather, career paths must be in place which retain and reward the increasing competence in their chosen, and presumably valuable, specialism. Our industry is seriously short of quality engineers, highly skilled specialists in particular aspects of the business. The shortfall cannot continue to be filled by graduates-in-transit to CEng, if the business is to become more efficient, effective and competitive.

So should employers be recruiting as many Honours civil engineering graduates in the first place? Should there be many more persons qualified with IEng-compatible degrees? Only if the industry requires them, and offers them a progressive career structure, will the universities respond and offer appropriate graduates.

Would it not, perhaps, be more cost-effective to recruit computer-literate and numerate 17-year-olds and train them for some of the specialist roles such as, for example, computer assisted design? Should employers be seeking and developing latent talent in their existing workforce? These trends are already apparent in many other professions, from banking and management consultancies to the police service and retailing. Why should engineering only be done by engineers? My solicitor does not write my will, nor does my accountant do my accounts, even though both retain the responsibility and ensure that it is done correctly. In any case, the developing law on age discrimination is making it more difficult to concentrate solely on graduate recruitment to the detriment of other age groups.

Should any of these people turn out to be capable of developing further than TMICE, then a variety of routes is already in place for them to become professional engineers at either IEngMICE or CEngMICE, wherever they started from (see Chapter 4).

Education

Many universities responded to pressure from employers seeking vocational competence, by producing graduates who could apply established knowledge and techniques to conventional problems, without developing any real understanding of the underlying principles of engineering. One of the main faults of unsuccessful candidates at both Incorporated and Chartered Reviews is still the failure to demonstrate their 'application of engineering principles'. They can use sophisticated software, but appear not to understand, or indeed question, the appropriateness of the underlying principles for their particular engineering problems.

Several universities which succumbed to the temptation to produce technical analysts, for the immediate needs of the business, are amongst those which have had to close their civil engineering departments, consequent upon the Engineering Council (EC) enforcing its new standards in 1997. So the supply of graduates content with purely technical roles is already diminished, forcing employers to rethink their recruitment strategy.

At the same time, other universities are moving in the opposite direction, increasingly recruiting excellent students and stretching them, during challenging and innovative Masters' courses, to develop the skills and understanding they will need to tackle the major social and environmental tasks this profession faces. These graduates will not tolerate an employer who expects them to spend an inordinate part of their early Initial Professional Development doing routine technical work, but will seek to build on their substantial education very quickly.

Chapter 4

Range of membership routes

The emphasis on strong mathematical ability as a prerequisite of studying engineering, in sharp contrast to the decline in numerate skills in secondary education, has led to a dichotomy, indeed perhaps a crisis, in engineering student recruitment. The inevitable result has been a whole host of relevant degrees, some equally as intellectually challenging as engineering, but without the emphasis on mathematics. These degrees in, for example, construction engineering or environmental management, economics, statistics or geography are unacceptable to the Engineering Council (EC) without significant top-up, because of their perceived lack of numerical rigour. But they can form a good basis for some of the roles undertaken by modern civil engineers.

Flexible academic base

In the past, the only route to qualification with ICE was to make good the educational deficit in some way, to satisfy the overriding requirements of the Engineering Council, and I have in the past guided many candidates down this route.

Now that the emphasis of professional qualification has veered away from what a candidate has done, to whether the candidate satisfies all the defined attributes (product not process), the Institution is able to widen its membership by inviting these graduates to become members without necessarily satisfying the EC's criteria (i.e. they cannot use the designatory letters CEng or IEng).

Associate Member

The Institution now offers several means of qualifying for membership, for those without accredited first degrees.

❏ If you are one of those who wish to remain as experts in their chosen (usually broadly scientific) field, with a three-year full-time (England and Wales) Bachelor's degree, providing specialist scientific or technical advice to civil engineers, you will be expected to demonstrate the same attributes as at the Member Professional Review, but with 'engineering' principles in attributes 1 and 2 replaced by 'scientific' or 'technical' principles as appropriate. You will be awarded the title Associate Member, with the designatory letters AMICE.

❏ Where your wider experience and personal desire has enabled you to apply the scientific or technical knowledge, obtained in your three-year full-time Honours degree, in the context of social and economic criteria, to resolve complex engineering problems (which is after all, the definition of civil engineering) then you may apply for the Member Professional Review. You will be expected to demonstrate how your experience has enabled you to develop the ability to make judgements on conflicting aspects of multifaceted problems, just the same as a candidate with an Honours degree in civil engineering, but in your own specialist field. You will be awarded the title Member, with the designatory letters MICE (but not IEng or CEng).

For those who do not have an acceptable academic qualification like those just mentioned, the Institution offers a number of means by which whatever qualifications you have can be topped up to satisfy the benchmark requirements. I strongly advise you to contact the Institution before embarking on any of these alternatives, where staff will be able to give up-to-the-minute guidance and help through an Academic Assessment.

❏ If you have an Honours degree in civil engineering, an MSc in a relevant subject will make it equivalent to the EC Master's benchmark for Chartered Engineer.

❏ If you have an Honours degree in any other subject, a civil engineering subject MSc will make it equivalent to the EC benchmark for Chartered Engineer.

❏ There does not, at the time of writing, appear to be a similar system to top up to the Member IEng benchmark, but no doubt, if demand warrants it, guidelines will be produced.

❏ For any other degree or academic qualification, the Institution will identify the shortfalls and offer suggestions as to how these could be rectified by Further Learning, which could be either, or a combination of, further academic study and/or workplace experience.

❏ If your original academic qualifications are not relevant or partially satisfy the EC benchmarks, then it could be that your experience has compensated for the shortfall. If so, and you have progressed to a position of responsibility commensurate with a professionally qualified engineer, there is yet another route to membership: the Technical Report Route. This route, which is a logical development of what used to be the Mature Candidate Route, is covered fully in Chapter 21.

Conclusion

The Engineering Council, consistent with the General Professional Directive of the European Union and its derivatives, seems set to continue to insist on there being two classes of professional engineer. The larger group, as in all other countries in the European Union, must be those more technical engineers (Ingénieur Technique), defined in the UK as Incorporated Members.

The construction industry, having resisted this split for many years, must now move swiftly to recognise and implement the policy, if it is to comply with European law, particularly the General Professional Directive and derivatives, as well as the requirements of the statutory regulatory body for UK engineering – the Engineering Council.

However, the Institution realises that many people make significant and substantial contributions to civil engineering, without necessarily having academic qualifications which satisfy the Engineering Council. It has therefore devised new, or expanded existing, routes to enable such persons to be professionally recognised and accredited.

Chapter 5

The jigsaw concept
for review preparation

Helping my young grandson to build a jigsaw, I realised that we were doing exactly what every candidate for review must do – construct a jigsaw. I was explaining to him that it was no good trying to find a piece with the correct shape; what we had to do was find a piece which added to the picture. Each piece had to be examined in detail to see how the part picture on it interlocked with the bits of picture on the pieces alongside. Continual reference had to be made to the picture on the box to see the position of each piece in the complete picture. It was no good picking up pieces which looked as though they might be the correct shape, without looking at the picture on the box. This process mirrors precisely the system which I recommend for compilation of your review.

All the components of the review, the documents making up the submission, the interview and the Written Test, are each an essential part of the whole picture. They form the pieces of your jigsaw, but the size and shape of each piece will vary from candidate to candidate. All the pieces are important; none can be omitted. Everyone remembers how frustrating it is when one final piece of a jigsaw is missing or is a bit mangled; you must not leave any possibility that the Reviewers might feel that frustration!

The picture is similar for everybody – either MICE, AMICE or TMICE, as indicated here:

But the pieces are different shapes and sizes for each individual – no two jigsaws will ever have the same shape or size of pieces.

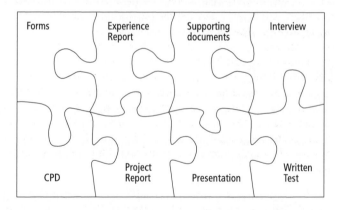

This is why it is no good copying someone else's submission – it probably will not work *for you*! It is also why the Institution is very reluctant to publish examples of successful work, either for the submission or in the written tests; there is always a great danger that any published example will be perceived as the approved format – the 'formula for success'. This danger exists within organisations as well, where candidates are tempted to copy an earlier successful submission in the belief that this is the way to succeed – many have had a nasty shock!

What you must do is ensure that *your* picture is complete, with each piece locked into the others and with no gaps or overlaps anywhere. It worries me when I am told, 'I've finished my Experience Report, now I'm going to start on my Project Report'. Just as, with a real jigsaw, we compile different groups of pieces as we recognise them, so in my view it is good tactics to develop all the pieces of the jigsaw together, so that they are each complete in themselves, and form a complete whole, interlocking without any overlaps.

The jigsaws for each class of membership have the same number of pieces, with slight variations in detail. For example, Technician Members do not have to undertake a written test.

Obviously, in the early preparation of your submission, you will not know what is going to happen at the interview, but you should, even at this early stage, have in mind what you might use for your presentation and how you might respond to the invitation to 'take us through (some other part of your submission)', informally at some point during the interview.

No Training Record

Your Training Record may have taken many years to compile, representing many years of personal development. At first, it may appear perverse (perhaps even slightly annoying) that the Reviewers do not see any of the resulting paperwork. You have after all, put a huge amount of effort into compiling it. But the Institution has devolved the responsibility for ensuring compliance with the requirements of Initial Professional Development to your Supervising Civil Engineer (SCE) (if under Agreement) or to a Career Appraisal if not. Keep in mind the overriding shift of emphasis:

> **Have you, as a result of what you have experienced, developed the attributes of a professional engineer?**

This fundamental change was the final step in a long process, switching the emphasis at the Review away from 'How did you become a professional engineer?' to 'Have you become one?' This removes undue reliance on any training system or quality of experience, and places the emphasis very firmly on the outcome, in common with the Engineering Council, which requires 'an increased focus on output standards' (*EC^{UK} SPEC2004*). The Reviewers will

review the product of training, not the process of training.

What all that training documentation ought to have done, however, is develop the skills, knowledge and understanding you now need to be successful, as well as developing a methodology and attitude which will stand you in good stead as you now set about preparing the submission documents for the review.

Subsequent chapters of this book consider each piece of the jigsaw in turn and how it interlocks with others, but the next chapter spends some time looking at the completed jigsaw – the picture on the lid of the box containing all the pieces.

Time to prepare for the review?

The professional review is the culmination of many years' work and personal development, which started long before you even went to university or college. It is the point at which many strands of development are brought together – technical knowledge, professional responsibility, personal characteristics – to demonstrate and prove your competence to fulfil the role as defined earlier, within your particular field of expertise. It is most certainly not an examination of whether you are capable of doing your job, which is unfortunately how many candidates (and some of their sponsors) seem to view it.

So one answer to the question 'How long will it take to prepare?' is 'Many years'. In reality, however, I hope you are continually reviewing your progress against defined targets – initially, your training objectives and, latterly, the available descriptions of what has to be demonstrated at the review.

There will come a point (usually quite suddenly) at which you feel confident about your ability – that is the moment to start preparing the submission. From this moment, and based on watching many candidates, I recommend a minimum period of some six months before the submission date. On this basis, a rough programme would be as follows.

Submission by	Draft reports by	Support documents by	Sponsors approached
15 February	end September	end January	end November
15 August	end March	end July	end May
30 June (overseas)	mid February	mid June	mid April

In summary, if you are submitting in the UK, you should be commencing preparation in earnest very shortly after the submission dates for the previous review session.

Such a programme provides adequate leeway for the inevitable delays in responses from your sponsors and advisers and for the Christmas and summer holidays. But it still means that preparation has to be pursued diligently and constantly if the submission is to be of the highest standard, demonstrating your capabilities to best advantage. I have known successful candidates who started long after the dates suggested above, but they virtually abandoned everything else in a frantic surge of 'accelerated working'!

In addition to the preparation of the submission, you will also have to programme into your schedule the preparation for the interview, including a presentation, and the written work. It seems that few young engineers encompass these wider issues in the normal course of their work or by research and reading around their profession, so it is usually tackled like some kind of crash course at the end. I deprecate this, but acknowledge that this is the reality of most situations. However, do not underestimate the work involved. It is unlikely to be successfully squeezed into the decreasing period between the submission and the interview.

Chapter 6

The complete picture

Before examining the detail of the 'picture on the box', I think it is important to consider the underlying purpose of the reviews. This is not some sort of one-stop examination, for which you can cram and swot and then forget what you have memorised a day later. It is a *review* of the benefits you have gained from the experience you have had and an assessment of whether these have developed the required understanding and skills – it is

a review of what you have become, not of what you have done.

Obviously, one cannot truly be separated from the other, but the focus is very definitely on you, not your work. Your experience and specific projects are the vehicles by which you demonstrate your abilities; they will not, in themselves, impress the Reviewers. There is no automatic correlation between the prestige or complexity of the work and the benefits you gained. Some people can have excellent experience but do not take full advantage of it. At the other extreme, some people have access to limited experience, but make maximum use of it to develop the requisite knowledge, skills and understanding.

Despite the best efforts of the Institution, there still appears to be widespread misunderstanding, exemplified by comments I continue to hear, about the purpose of the reviews. There is nothing mystical about the reviews; the present system is the culmination of well over

60 years' experience in the review process. Further changes will be made as more operational experience is gained and as the competencies being reviewed become more clearly defined. However, three problems will always be inherent.

(a) The personal qualities and understanding being sought are intangible and therefore difficult to define.

This is the reason why there is an almost continual complaint that the Institution keeps 'moving the goalposts'. I do not think it does! Fundamentally, I believe that the Institution was seeking substantially the same abilities and understanding 35 years ago, when I submitted, as it is today, even though the descriptor words may be different. Engineering is like 'management'; if there were only one agreed definition for all time, there would not be so many books on it! Neither are the requirements static, but shift as the demands upon our profession change.

(b) Young engineers dominated by (and highly proficient in) very structured examination systems during their formal education, will always seek syllabuses, model submissions and defined targets. None of these can be produced.

Candidates familiar with comprehensive academic syllabuses, are flummoxed by the lack of guidance and rules for the Review. The Institution offers a general statement of what must be demonstrated, but little or no guidance on how to demonstrate it – it is a performance specification rather than a method specification.

So candidates seek precedents; examples of what others have done to be successful. Hence the creation of all the myths about what experience you need, and what the Reviewers expect, all of which may be well-meaning, but are misguided and restrictive. For the same reasons, many candidates try to emulate someone else's successful submission – it won't work, because every individual engineer is different and their experience is different, even if they are working on the same projects.

(c) This is a review, not an examination! A review of your capabilities, some of which you may not yet have had an opportunity to utilise.

At the time of the review, candidates are in a 'chicken and egg' situation – 'I believe I have become a professional engineer, but I am not doing a professional engineer's job yet because I am not qualified as a professional engineer'! So to some extent, the Reviewers are measuring potential. The extent of that will obviously vary with age, available experience and responsibility to date, and requires the Reviewers to make a judgement. It is a judgement of the whole person, your capabilities and understanding, not a collection of disparate details; what is known as the holistic review.

The holistic review

The review process is described as 'holistic', a word which does not yet appear in every dictionary, but is more usually associated with alternative medicine and ecology. It comes from 'holism', the belief that the whole is greater than the sum of its parts. The Reviewers are encouraged to concern themselves with the whole of your capability, rather than analysing each separate part in detail. Their training emphasises the interdependence of the various attributes being sought, in the context of your particular work environment, and the need to make a judgement about your overall capability. Someone who trains examiners for the UK driving test says that the examiners are told to sit through the test, and decide at the end whether the person is safe or not. If that decision is 'No', then they have to think through why they made that 'instinctive' judgement and list the reasons. A very similar methodology is used by your Reviewers.

There is no weighting; one part of the review is not more important than any other; no one attribute takes precedence over any other.

This means that it is possible (and it *has* actually happened) that whilst one aspect may have been found less than acceptable, the person has satisfied the Reviewers that, overall, they are perfectly capable of discharging the responsibilities of a Member in their particular circumstances. Just as I stalled the car during my driving test, but still passed! However, I would not recommend that you rely on this rarity; make sure you demonstrate that you are at least adequate in every respect!

The fundamental problems of the submission

As a result of guiding over 2000 candidates through the reviews, I now believe that there are fundamental problems inherent in the whole process. These difficulties need to be addressed by every candidate as they start to think about preparation of a submission.

Personal opinions

For years engineers, even during science lessons back at school, have been taught to write without reference to the writer: 'Such-and-such was decided' or 'It was agreed that...'. Now you are being required to write in the first person: 'I decided' or 'In my opinion, it might have been better if...'. The reviews require you to demonstrate that you have become a Professional Engineer, not just list the experience you have had. So you *must* write about yourself and your personal opinions, which goes against everything you have been taught. Engineers do not like writing about themselves, particularly for the scrutiny of two strangers; we find it embarrassing. Do not underestimate this problem; it is a significant hurdle in the preparation of a good submission.

Self-confidence

Civil engineers are in the business of risk management, which is always a matter of judgement, and so do not boast. We are rarely, if ever, certain that our decisions are absolute. How can they be, when the parameters are constantly changing? We like to talk through problems and solutions to reaffirm that we have solved them correctly – there is a huge potential benefit, if taken advantage of, for entrants to our profession to learn by discussion. If confronted (for example, in a court of law) we rarely if ever say 'Yes' or 'No', without some qualification – 'Yes, perhaps...' or 'No, but...'. Now you are required to show that you are decisive and able to inspire confidence in others.

Owning responsibility

We all know that we do not work in isolation, but are part of teams. So we are loathe to take credit for our personal decisions, knowing that they have usually been discussed with many people before implementation. The result is that we write 'We decided' or 'A decision was made'. But who would take the blame if subsequently these decisions were found to be incorrect? If it is you then, no matter how many people you discussed it with, it is *your* decision – 'I decided'!

Assessing others' decisions

By and large, at the time you present for review, you are still working at a relatively low level in the hierarchy. In many organisations, and certainly in some ethnic cultures, it is not acceptable, if you wish to progress, to openly discuss decisions taken by more senior engineers. But the Reviewers will *expect* you to have views on how high-level decisions were made. In the future, you might be expected to take similar decisions as a qualified professional engineer, and you must demonstrate that you could.

Demonstrating understanding

Most people coming up to professional review have progressed through many years of education – at school, college and/or university. They have developed many techniques to successfully negotiate detailed syllabuses, coursework and, particularly, examinations. If you attempt to use the same techniques for the professional review, they probably will not work. The review is not an examination! It is a review of whether you have developed enough skill and understanding to operate as a professional engineer. Obviously these are dependent on some knowledge but, unlike most examinations, the review is not primarily a test of knowledge but a test of your understanding. The Reviewers are continually asking the question, 'Would this person, when placed in a position of responsibility, make the right decisions?' Note 'would' they, not 'have' they. In other words you may not yet have had the opportunities to put your capability into practice. This concept of capability is at first difficult to comprehend, but must be realised if a successful review is to be the outcome.

Chapter 7

Professional engineer

I included four 'definitions' of a Chartered Engineer and three of an Incorporated Engineer in my book on effective training; all are widely available in publications from the Engineering Council, the Fédération Européene d'Associations Nationales des Ingénieurs and our Institution (the Chilver Report). All date back to the 1970s, so there is nothing new in them – I merely brought them together. Council of the Institution published another set of descriptions in 1998. In fact, engineers have been attempting to 'define' themselves since long before Henry Palmer and Thomas Tredgold in 1828 – even Aristotle had a go!

We all remember Tredgold's 'the Art of directing the great sources of Power in Nature for the use and convenience of Man etc.' but how many recall that other part of our Charter, which calls for

'a high degree of professional knowledge and judgement in making the best use of scarce resources in care for the environment and in the interests of public health and safety'?

Yet every candidate professes to have achieved the Objective about the history, role and purpose of the Institution!

I now have a collection of 'definitions' from all over the world; all are trying to say the same thing, but none is a 'prescription'. If there were an easy definition, then all the prolific writers on management, let alone engineering management, would have long since run out of ideas!

The only readily available information is in the booklets *Routes to Membership ICE3000 series*. While these are by and large 'method' specifications for the review itself, they do include coverage of the 'performance' expected, in Appendix A, set out in a table.

This table encapsulates the whole specification for the reviews; every part of your effort so far should have had these attributes in mind – your experience, training, Continuing Professional Development (CPD). Even if, up to now, your understanding was a bit vague, preparation of the submission documents, as well as your preparation for the interview and any written test, must now focus exclusively on these attributes and how you developed them.

I believe the purpose of the reviews can be summarised in two sentences:

Do you understand what you are doing?
Do you understand the implications of what you are doing?

For a Technician Member, there is far less emphasis on the second of these sentences, while, at the other end of the spectrum, there is considerable emphasis on the wider understanding and full implications of your role if you are applying to become a Chartered Member.

Deciding whether you have become a professional engineer

It is evident that many engineers, both candidates and senior managers, have still not come to terms with the fact that the Member Professional Review encompasses the more technical part of the old Chartered Engineer spectrum. Chartered Professional Review candidates are still coming forward with an attitude and understanding of an Incorporated Engineer, despite both the Institution's and the Engineering Council's best efforts.

But worse, even now, I still occasionally hear that 'you need a design for your civils', that 'you need twelve months on site/in design', that 'you need a bill of quantities and a rate build-up', that 'you can only count civils meetings if you have written 500-word reports', that you need to perambulate around all the departments in an organisation before a Training Agreement can be signed off. You may have done some, any or all of these for your personal development; it might be a good idea, your SCE may insist, but they are not *requirements.*

There are very few 'rules' on how you develop.

The requirement is that, at the review, you

demonstrate that you have achieved the required skills and understanding.

What is required is that, by some means or other, you gain adequate experience to develop the characteristics being sought. The end product is 'defined' (albeit very loosely), but not the precise means of achieving it, apart from some very rudimentary objectives. Of themselves, achievement of these will not be enough, but they do form a solid foundation upon which 'responsible experience' can be built.

Work experience needed

I can assure you from watching many engineers progress successfully through the reviews, that it is possible to gain adequate experience in the most unexpected situations. No longer does the Institution have routes for so-called specialists; at last the reality is acknowledged, that

every candidate is a specialist,

whether in structural concrete, steel or masonry, bitumen, geo-textiles, groundwater movement, coastal protection, North Sea oil, traffic calming, public transport, even as far removed as insurance assessment and actuarial work – but all must be able to display the skills and understanding required.

There can be no specification for your personal development, because everybody starts from a different point, has different innate abilities, differing education and different experience (even if working in the same office).

Some trainees I meet never had the opportunity for a Training Agreement. During the deep recession in the early 1990s, for some, employment was a series of temporary contracts (some of only a few months' duration) with organisations not on the ICE Index of Employers Approved for Training. Yet they had exemplary Training Records. How? Because the purpose of training and the required end result were fully understood and senior people had been inveigled and cajoled to assist. Their records contained a complete set of critically annotated Quarterly Reports, a ratified set of achieved objectives and more than adequate confirmed Continuing Professional Development. Furthermore, all the required abilities were evident, both in their submission documents and at review.

These examples imply that all the information needed *is* available – but still the hearsay and half-truths remain. It is difficult to understand why, except that many people seem far more comfortable with a set of rules which can be ticked off as completed, thus avoiding the need for judgement and assessment against a series of rather vague criteria. What each individual must do is

continually assess yourself against the end-product,
identify any deficiencies
seek experience or off-the-job training
rectify any shortcomings.

Chapter 8

The review specification

Expansion of the *ICE3000 series* Appendix A tables

Now to return to those items in Appendix A. At first glance, they do not seem too onerous, but once you start trying to demonstrate your ability in each of them, they become far less clear. Like all specifications, they should be read with the intention of finding out how best to comply, i.e. achieve the performance, but this is not straightforward.

Personally, I find the closer alignment of the new Appendices A with the format and criteria laid down by the Engineering Council in their generic document EC^{UK} *SPEC2004*, is perhaps not as user-friendly for candidates as were the former paragraphs 14.2.1 in the *ICE2000 series* documents. But perhaps this is because the *ICE2000 series* was in existence for over 15 years and we developed 'case law' about what was meant, i.e. the interpretations became clearer with experience.

So it is useful to reflect on, perhaps anticipate, how the new criteria will be interpreted. I have taken each section in turn and amplified it in the context of other documents about the responsibilities of professional civil engineers, issued by the Institution and others. However, my interpretations are just that – my interpretations – and you as a candidate should think the problem through for yourself, hopefully with the cooperation of your mentors and by reading Institution policy as it emerges. The Institution has offered some more suggestions in their guidance notes. As experience becomes available, there will inevitably be greater certainty of interpretation, so check that you have the latest guidance note.

The following paragraphs are numbered and lettered to relate to the attributes set out in Appendix A for your particular class of membership. Each section includes MICE (IEng and CEng), MICE (no EC designation), AMICE and TMICE attributes, all of which are similar but with differing emphases.

1. Engineering (Technical) Knowledge & Understanding

Engineering Knowledge & Understanding

Engineering knowledge is an understanding of the fundamental properties of materials and the basic behaviour of engineering systems, i.e. the technical principles of strength of materials, soil mechanics, hydraulics, structures and the many other subjects covered during your formal education as a civil engineer. I, and my co-Reviewers, become concerned whenever, for example, a candidate is unable to draw simple bending moment or shear force diagrams for structures they have analysed, or to state the basic formulae upon which their calculations of flow rates in pipes are based. If you have used Young's modulus to check possible deflections, or software based on D'Arcy's law to estimate groundwater movements, then surely it is a prerequisite that you know what they are?

This understanding of the engineering principles underpinning technology provides the basis from which existing and emerging technology can be properly applied. There are two fundamental reasons for this requirement:

❏ without such fundamental knowledge, technology (particularly software) might easily be misapplied;

❏ it is imperative that the results from sophisticated programs are checked by some rudimentary calculations, quick design methods based on engineering principles, to make sure that the results are of the correct order of magnitude.

It is the proper understanding of engineering principles which enables you to develop the imagination to anticipate possible loading conditions, thus enabling you to make accurate judgements about suitability, risk, safety and the proper use of resources, which are the criteria behind the optimisation of application.

In addition, for the Chartered Professional Review, you must demonstrate that you are able to use your understanding of engineering principles as the basis for both developing and exploring new technology.

Technical Knowledge & Understanding

Much of what has been written above applies equally to technical and scientific knowledge. Both types of knowledge require an understanding of the fundamental properties of matter and the basic theory behind the processes of analysis you use; for example, statistical probability theory, the biological and chemical principles of remediation or decomposition, or any of the sciences upon which your work relies, probably covered to some extent during your formal education. If you have estimated groundwater movements, or made estimates of future traffic flows based on limited samples of the existing situation, or predicted the effects of an earthquake, then surely it is a prerequisite that you understand the principles behind your work?

This understanding of the technical principles ensures that existing and emerging technology is properly applied. Without such fundamental knowledge:

❏ technology (particularly software) might easily be misapplied;

❏ the results from sophisticated programs cannot be checked by some rudimentary calculations, quick methods based on technical principles, to make sure that the results are of the correct order of magnitude;

❏ it is unlikely that you will develop the imagination to anticipate possible scenarios or the understanding to make sensible judgements about suitability, risk and the proper use of resources.

2. Engineering (Technical) Application

On a superficial reading, this attribute group does seem to overlap with group 1. On this basis, it is the ability to apply established analytical methods and procedures appropriately, something which requires a thorough understanding of the principles upon which those procedures rely. However, as I read it, 'procedures' are not restricted to technical analysis, but would include all those systems used for the management and control of the enterprise.

Further consideration reveals that this group also covers the application of principles as part of the whole solution to a problem. So this needs the development of a wider perspective, where candidates can demonstrate that they are able to choose the optimum techniques, procedures and methods – what is often referred to as appropriate technology – fitting the means to the end. So, as a simple example, what might be appropriate in an environment where labour is cheap might not be suitable in another place where wages are relatively high. A sophisticated analytical technique producing a complex high-tech solution might not be appropriate in a newly developing country. Safe working conditions may be radically different from one country to another, dependent upon the perceived value of human life.

There is also a requirement to evaluate the effectiveness of solutions. To look back and gather that great gift – hindsight – which always arrives slightly too late! To watch the maintenance and operation of the solution and determine what might have been done better or more effectively. Only then can better solutions be devised in the future. So do not be surprised during your interview if a Reviewer asks questions like, 'Would you do it the same way again?'.

3. Management and Leadership

Management is guiding and directing functions, resources and people within the context of a given project or activity. There is, to me, a distinct difference between management ('guiding and directing') and supervision ('controlling'). Supervision is about telling people what to do and how to do it, leaving precious little space for them to use their own initiative. Good management requires the replacement of direct control with mutual trust, a difficult but rewarding transition for both employee and manager.

(A) Your own self-management needs development. Starting out in the workplace can seem like entering a never-ending avalanche, with no pause for thought. Work just pours in and you are expected to juggle several jobs at once. Your managers will just presume you are able to cope, but few new employees can, without some basic guidance and support.

Prioritising work, management of time, staying calm under pressure, are all things you have either learnt from hard experience or been taught, and you should demonstrate these capabilities by examples in your submission.

(B) Management is directing people towards the outcome which needs to be achieved and, perhaps, offering them guidance on how it might be accomplished. As you approach the Member Professional Review, you should no longer need to be supervised, and ought to be taking more responsibility, having developed the knowledge and confidence to accept it. You must demonstrate that you can manage yourself and others in your team effectively as well as being capable of obtaining the necessary resources (such as time, equipment and knowledge) for the team to achieve the required outcome. And all of this must be in the context of satisfying quality processes.

(D) To become a Chartered Member, you are required to be capable of leading activities and change. Leadership is defined by the ICE as being capable of 'setting the direction of a project or activity and encouraging and guiding people towards that direction'. This agrees with the *Leadership Trust* definition: 'Leadership is using one's own personal power to win the hearts and minds of people to achieve a common purpose.

Leadership is not a function of your position, but a desire to persuade people of the validity of a course of action which you envisage. It is as visible in a primary school playground as in higher management, and must be demonstrated to the Reviewers, even where you might consider that you are not yet in a position of leadership in the hierarchy of your organisation.

(E) At CPR, you must demonstrate that you have successfully challenged others to progress their abilities to meet new challenges, to work beyond areas when they have previously felt comfortable.

(F) For a Chartered Member, leadership is having a vision of what you want to change, where you want to improve, and inspiring others to that end. Leadership is about personality and a willingness to take calculated risks – being prepared to 'put yourself on the line', motivating and inspiring others to follow you.

The ideal employee straddles both management and leadership, but there is undoubtedly a clear difference of personality and effective zones of influence between the two. Neither is exclusive, indeed there is considerable overlap, but there is a distinct bias for each. One tends towards routine orderliness, the other towards disruptiveness – challenging the *status quo*. Which one best fits you?

(E) If you want to become a Chartered Member, you must record occasions where you have managed to persuade others of the validity of your proposals and inspired them to carry the solution through.

(F) You must give examples of situations where you have strived for improvements to the quality of service, for example, in the use of scarce materials or other resources, in care for the environment, in protecting the public interest, or in the efficiency and safety of the solution.

4. Independent judgement and responsibility

(A) You must demonstrate, by examples from your experience, that at times you have felt unsure of your ability, and have sought guidance from people with more experience. The responsibility for 'knowing when you don't know' is considerable and important and must be demonstrated to the Reviewers.

(B) You must also demonstrate your ability to go beyond established techniques, proven methods and documented precedents to

develop a solution to a problem and to take personal responsibility for the effectiveness of the solution. You need to
demonstrate to the Reviewers that, despite being well aware
of best practice as defined by Codes and Standards, there
have been circumstances where you were unable to apply it,
but had to develop a compromise which, in your judgement,
would work satisfactorily. That means that you took responsibility for the decision, and its possible outcomes, and are
expected to be able to justify it.

(C) For the Chartered Professional Review, you are required to
demonstrate that, not only can you identify the limits of
your own personal knowledge and skills, but also those of
the team around you. This is a fundamental management
challenge: to delegate work which your staff feel confident
that they can achieve, but which at the same time presents a
challenge – pushing them outside their comfort zone to
develop their abilities still further.

Even where you might at first consider that you work on your
own, there will inevitably be times when you are dependent
on the input of others (e.g. surveys, technical or procedural
expertise, marketing) and you have to make judgements on
the reliability of those people and the validity of their advice.
Explain how you did it!

(D) Similarly, as well as demonstrating engineering judgement,
potential Chartered Engineers must demonstrate that they
are capable of making wider judgements beyond solely
technical matters, to achieve fine balances between such
things as:
❑ financial costs and commercial benefits;
❑ damage to, and benefits for, the environment;
❑ capital investment and maintenance costs;
❑ local disruption and wider social benefits.

5. Commercial ability

(A) Budgets are the means by which the use, control and documentation of expenditure are achieved. Competence to prepare a
budget requires you to be able to estimate the costs of doing

work, the profit which needs to be achieved and, probably, how to optimise cash flow.

Budgetary control in most organisations is subject to in-house procedures, and you should know these thoroughly and demonstrate that you are confident, not only to apply them, but to use them as guidance where the particular situation is not exactly covered by the procedures.

(B) Statutory and commercial frameworks – I presume that this section includes the whole legal framework within which we are required to operate, not just commercially, but technically, environmentally and socially. I consider it absolutely vital that everyone working in risk management, which is arguably what civil engineering is all about, fully understands their liabilities and duties in law. There is a plethora of new legislation, which as yet shows little sign of abating, as politicians mistakenly respond to the growing blame and compensation cultures.

An early understanding, and continual updating, need positive action by both yourself and your employer, probably by formal training, since it could easily be too late (and expensive – or worse!) if such understanding is picked up by experience. At the time of writing, construction health and safety do seem to be targets for this sort of critical attention, but this emphasis is not often mirrored in the many equally crucial aspects of our work. Environmental and social impact, and effective use of resources, are just as important.

(C) Chartered Engineers must demonstrate that they understand the balancing of costs and benefits in a professional manner to protect and further the interests of those to whom the organisation is responsible (e.g. clients, subcontractors, general public, staff, shareholders). They should know a little about such things as equity, working capital management, forecasting and presentation of accounts (profit and loss, balance sheet and cash flow) of the organisation.

6. Health, safety and welfare

I am concerned by some of the young engineers I meet who seem to think that health and safety (welfare is rarely even mentioned) is all about such basic things as personal protective clothing, toe boards and risk assessment form filling. The emphasis on such matters by the Health and Safety Executive and others, combined with combating an increasingly litigious attitude in society, has inevitably drawn attention away from the fundamental concepts to the details of compliance with rules. It is similar to treating the symptoms of an illness, rather than the illness itself.

I am pleased that recent publications from the Institution are re-inforcing a 'back to basics' approach. To me, the health, safety and welfare of everyone, not just constructors, has always been the bedrock upon which all my decisions were made. They consti-tute an attitude of mind, well beyond the humdrum details. The increasingly 'knee-jerk' reactions of politicians and the public to accidents can lead to a reactive and grossly distorted waste of resources, if not tempered by the logical and pragmatic risk assess-ment of engineers. For example, was the widespread increase in the length of motorway safety barriers justified because one driver fell asleep and drove on to a railway line? Was that the best risk mitigation or avoidance?

Forms and procedures are the means by which this fundamental attitude and the process of assessment are recorded and commu-nicated. Systems must never be seen as an end in themselves, but as a means by which responsibilities are discharged.

Health

Who better than the World Health Organisation to tell us what health is and what is involved in achieving it:

> 'health comprises those aspects of human health, including quality of life, that are determined by physical, chemical, bio-logical, social and psychosocial factors in the environment'.

This is a much wider description than most of the engineers I meet seem to realise. It is a bit wordy, but it does go on to spell out our overall responsibilities, and the need for judgements, very clearly:

> 'It is the theory and practice of assessing, correcting, controlling and preventing those factors in the environment that can poten-tially affect adversely the health of present and future generations'.

Safety

What about safety? There is an attitude apparent in large sections of the public that somehow all risk can be avoided and if it is not, then someone else must be to blame – the 'compensation culture'. The news talks of risk aversion, both in schoolchildren and in business. There are people who truly believe that the railways should never have a derailment, or that there should never be accidents on motorways. People are concerned about the risks of travelling by air or rail, but think nothing of driving on motorways which have a far greater accident risk, killing 3000 people annually. Public perception is a fickle thing, and we must avoid being sucked into irrational reactions.

Safety is about *managing* risks, not trying to eliminate them, which is a lost cause anyway. The Health and Safety Executive states (2005) that

> 'the Health and Safety at Work Act does not require firms to obey inflexible, hard and fast rules, rather they need to assess the risks that result from the work and identify sensible control measures that are proportionate to the risks'.

So I believe that all candidates must be familiar with the anticipation and management of hazards and risks in investigation, design, construction, use, maintenance and demolition, within criteria defined in law or otherwise established as best practice. Obviously, the bias will be towards those parts of civil engineering in which you have direct involvement. While for a potential Technician Member, it may well be mainly about compliance with procedures and systems, as a potential Member you are required to demonstrate that you have used your judgement in achieving acceptable situations and are prepared to defend your decision, not merely comply with rules in an attempt to avoid blame and possible prosecution.

Welfare

This is a more nebulous concept, the 'faring well', or well-being, of everyone involved, not just constructors, but the users, maintainers and demolishers of whatever infrastructure is being developed. The benefits cannot so easily be demonstrated. Safety can perhaps be 'measured' (or the lack of it) by a reduction in accidents, health by changes in the number of days off work, but the benefits of welfare are not so clear. Yet all three are interrelated.

Welfare overlaps considerably with health and safety, particularly when there is a serious attempt to develop a 'safety culture'. The provision of adequate and decent washrooms, toilets and catering facilities, well decorated and spacious offices or comfortable safety and weatherproof wear which allows freedom of movement, do not of themselves improve health and safety, but they do help to create an attitude. Offices with few, if any, external windows, no views and inadequate lighting and ventilation, do tend to have low morale, higher staff turnover and sickness, contributing to the term 'sick building syndrome'. Those who have experienced the noise of working next to highway contra-flows will understand the sanctuary of soundproofed accommodation, not itself a safety measure, but surely contributing something to a safer working environment by reducing stress?

Should we still expect manual workers to go to and from work in the same clothes they use for work? Indeed, should washing and clothes storage places also be provided for those who exert them-selves getting to work – cyclists and joggers? I certainly found a huge jump in morale and an improved attitude of responsibility when washing and changing facilities were provided for the staff on all our waste disposal sites. There were tangible improvements in health and safety, which showed themselves in less absenteeism and fewer recorded injuries. If the employer is seen to care, then it appears that the employees will respond favourably.

So do not believe that welfare is restricted to site work. It affects all aspects of infrastructure development, use, maintenance and demolition. Take note of what is being done around you, what you could do or suggest to provide a better work environment, and demonstrate your concern to your reviewers.

Summary

As stated earlier, in a nutshell,

do you fully understand the consequences of your work?

Those applying for Chartered Membership must also demonstrate that they are positively changing the attitudes of others, that they are using their vision and initiative to strive for continuous improvements in health, safety and welfare performance. Again, this does not necessarily require you to be in a position of authority; we can all influence attitudes and behaviour by example, or appropriate discussion and criticism. Show that you have!

7. Sustainable development

Sustainable development is the pragmatic response to the concept of sustainability. Sustainable development is the management of resources in a project to maximise the benefit while minimising the disbenefits to the environment. For this purpose, the definition of the environment is

> **the aggregate of all the external conditions and influences affecting all forms of life on this planet, both now and in the future.**

The idealistic concept of sustainability, 'satisfying the needs of today without compromising the ability of future generations to meet their needs', seems to suggest that sustainability only applies to the human race ('future generations') but I feel sure this is not what was intended by the World Commission. Hence the broad description of the environment in the preceding paragraph. In any case, ideal sustainability cannot be realised all the time we are, for example, using fossil fuels, quarrying stone or making steel, cement and bricks. But we can, and must, work towards the ideal, and everyone approaching the reviews must demonstrate their awareness of current best practice and how it is implemented in their work. The Institution has produced guidance on this, and several other issues; have you read it and put it into practice? If so, tell the Reviewers about it. Candidates for the Chartered Review must additionally demonstrate that they are continually seeking every opportunity to make inroads into further reducing our profligate use of resources.

A sound knowledge of and commitment to sustainable development can be demonstrated by showing that you exploit any chances to reduce the use of finite resources, through such everyday measures as:

❑ fitness for purpose – neither grandiose nor over-demanding over the life cycle;

❑ reclamation, recycling and reuse;

❑ reducing specifications to allow secondary materials to be used where safety and fitness for purpose are not compromised;

❑ refit rather than rebuild;

❑ energy conservation in the whole life cycle.

8. Communication

The criteria are pretty clear regarding what the Reviewers expect candidates to demonstrate, but perhaps it is worth considering how those abilities may be achieved?

Engineers must gather, absorb, assess and process information (GAAP) and communicate the result (bridge the GAAP) in a manner which is understood by any of a wide variety of recipients, from operatives to politicians, from pressure groups to colleagues, and most importantly of all, the public. The continuing suspicion of many of these groups and continuing misunderstandings between colleagues, do suggest that we are not particularly good at it. For example,

❑ How many changes in your work have resulted directly from misunderstandings or lack of information in-house?

❑ If the general public is overwhelmingly hostile to a proposal, then surely it is our fault; we have not explained the reasoning behind our proposals adequately.

❑ Any credibility which a pressure group has must be as a result of loopholes in our information supply (or we've made a mistake!).

I know we are often confined by political considerations, but in general I do not believe we take the public into our confidence early enough or comprehensively enough.

Communication is about *information.*

The efficiency of communication is dependent upon the *effective* relaying of *just enough* information to achieve the desired effect on the receiver.

The quality of the information and the method of delivery will determine the level of performance by the receiver.

Information is not just facts, but also covers things such as goals and targets, policy, support, encouragement, discipline, attitudes, even dreams.

We now have so many easy, instant methods of communication that I fear that perversely, they have actually become an obstacle to effective communication. Our Victorian predecessors managed to build mammoth projects all over the world using stagecoaches, sailing ships and slow, overland postal services – because they communicated well. The limited and lengthy methods available forced them to consider very carefully what needed to be sent to whom, when and how.

Today, all of us suffer from information overload; everyone sends everything to everyone else without any real thought, and seriously believes that they have communicated! Software, e-mail, the internet, satellite links are only tools; they cannot add to the basic message and unfortunately encourage too much information to be sent to too many people. As a result, we give it all a cursory glance and delete most of it, including sometimes, key matters which are hidden away in a mass of irrelevant words.

Instant delivery also has the insidious effect of requiring an instant response. The speed of delivery may be instant, but the speed of human thought has not changed for generations! While the sender may expect an instant reply, pause, take time to consider the ramifications of their questions and your possible answers. It would not be the first time that an instant response has resulted in a commitment to huge additional, and unbudgeted, expenditure!

Effective communication is achieved by the sender, who must consider all of the following.

❏ What is the required outcome?

❏ What the message needs to be. What does the recipient need to know in order for the sender to achieve the desired outcome?

Just enough information needs to be sent to ensure cooperation and response.

❏ What method of transmission is best?

 ❏ Speaking (face to face, telephone or video-conference)?

 ❏ Writing (letter, report, PR release, exhibition, e-mail, fax)?

 ❏ Visual (pictures, drawings, models, body language)?

❏ When is the best time to send the message?

❏ What language will the recipient understand?

❏ In what context will the message be received?

 ❏ How much does the receiver know already?

 ❏ What mood will they be in on receipt?

 ❏ What attitude will they have on receipt?

❏ How will the sender know that the message has been received successfully, i.e. that the receiver knows what the outcome must be?

Few people think these important criteria through before firing the message away! If they did, then the computer industry in particular would notice a significant downturn in traffic. Already some firms are preventing e-mail replies being sent for 24 hours, and disabling the clustering of addresses, to force their staff to think about the necessary content and key recipients before responding.

The other contributor is the receiver, who must listen (or read). In today's frenetic society, this skill is being lost; the art of not only listening to what the sender is saying, but deciphering what they are trying to say. Most people are poor communicators, i.e. they send too much information in an unstructured way, hoping the receiver will be able to pick out what they need. Because receivers suffer from information overload and the structure is garbled, the message is lost.

I have experimented with the television weather forecast, videoing it and then asking people what the weather is going to be in their area immediately after the presenter has finished. It is amazing what answers I am given, and the resulting embarrassment when I play back the video and, for the first time, the viewers actually listen. The sophisticated computer graphics available, the apparent need for an animated personality, the desire to justify their predictions by showing what is going on out in the Atlantic, all actually obscure the message. The weather forecast has perhaps become entertainment rather than clear, precise information?

You will, almost inevitably, find it difficult to convey to your Reviewers 'just enough'. The temptation will be to tell them everything, in the hope that they can pick out what is relevant. Most of you will, in my experience, start with reports which are far too long, and have to reduce them significantly to keep within the word count. In the chapters on the reports I suggest ways in which you can avoid this chore. Remember – you must *demonstrate* your capabilities, the Reviewers are not required to seek them out. One of those capabilities is your ability to get important messages across clearly and succinctly.

You may perhaps believe that quality assurance procedures help to ensure good communications. In my experience, quality assurance is a control system for the *process* of relaying information, providing an audit trail of what was done and said, by whom, and why. But it does not control or measure the *efficiency* of communication, only that it happened.

9. Professional commitment

Too often, in my experience, this has been narrowly interpreted as accumulating sufficient Continuing Professional Development (CPD) days and attending the odd 'civils' meeting. In my view, it means far more than that. It is, as stated by the Institution,

> 'the exercise of professional skill and judgement with integrity and to the best of our ability to safeguard the public interest in matters of safety, public health and the environment and to uphold the dignity and reputation of our profession'.

Professional commitment is about ethics, responsibility and morals. You will inadvertently reveal your true attitudes during the review, as you answer questions and express opinions. So do make sure you have read, understood and put into practice, the Institution's guidance on ethics and conduct, so that it has become an intrinsic part of your attitude to life and work.

Summary

All these review criteria enlarge on the Institution's Royal Charter, which all candidates profess to have read and understood to satisfy the Development Objectives:

> 'a profession which calls for a high degree of professional knowledge and *judgement* in
> > making the best use of scarce resources
> > care for the environment and
> > in the interests of public health and safety'.

The keyword implicit throughout all these descriptions is *judgement*. It will not be enough to demonstrate that you always comply with 'the rules'. The rules must inevitably be out of date, based as they are on previous experience, and there will be occasions when they cannot be applied sensibly. That is when your judgement comes into play. A Queen's Counsel once gave me a fundamental explanation of the basic thought process we go through for every decision:

> 'At this time, with these resources, in these circumstances, this, in my considered opinion, is the best solution'.

Having made that judgement, the decision can be defended as correct in a court of law. This does not mean that this particular solution is absolute, correct for all time, since any of those factors can, and probably will, change in the future.

Chapter 9

Difference between a Member (or Associate) and a Chartered Member

There is a clear distinction between the attributes of a Chartered Member and those of a Member or Associate. This was clearly spelt out by Council of the Institution in 1998:

'A Member (or Associate) is an expert in their particular field. Their role demands a practical approach, considerable technical (or scientific) competence and some managerial expertise and control in a particular aspect of the construction industry, with an understanding of the whole procurement process and an appreciation of the social, economic and environmental impact of their involvement.

A Chartered Member, on the other hand, combines a thorough understanding of technical principles with broad, multi-disciplinary professional and leadership capability to enable them to eectively and safely direct, change and progress the infrastructure and built environment by balancing financial, social, environmental and political implications and the eective and beneficial management of resources.'

Generally, engineers do seem to divide into different kinds of people. Some are at their best 'doing' the technical engineering; they consider that persuading politicians and a sceptical public of the legitimacy of their proposals or seeking permissions and managing resources is frustrating, or they may not, so far, have had the opportunities to operate at these interfaces. Others revel in the persuasion and conflict of trying to convince the public that what is being proposed is the best compromise solution, or of working in and organising a multidisciplinary environment, but retain sufficient technical understanding to realise the full implications of what they are doing.

At the Member and Associate Professional Reviews, you must exhibit a high standard of expertise in a limited range of work, where you could, if necessary, supervise others doing similar tasks. But you are also expected to understand the whole context of your work. So, dependent on your particular area of expertise, your profile might look something like this:

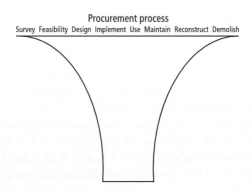

The area of greatest expertise will clearly swing from one extreme to the other. Someone working in traffic surveys, for example, will swing well to the start of the procurement process, someone in demolition will swing to the right, the end of the process.

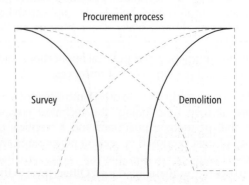

As a candidate for Chartered Membership, you are expected to have developed a greater depth of understanding of the whole procurement process, gained partly by reading widely and partly by experience. So you may not perhaps have such a depth of expertise in any particular specialisation; your profile would be quite different and might look something like this:

Procurement process

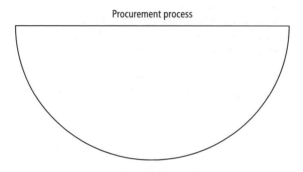

For completeness, the profile of a Technician Member is even more pronounced, where they know a great deal about one particular aspect of the procurement process, and relatively little about the whole process apart from how they interface with their immediate colleagues.

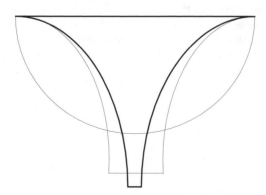

Obviously these profiles are a gross oversimplification of reality. They do in a sense divide the world into racehorses and carthorses – both good at their job but capable of doing the other job, perhaps not particularly well (although I quite fancy the idea of a carthorse race!). Where do eventers, steeplechasers or hurdlers fit? Somewhere in the envelopes below? Even more difficult to place is the highly developed dressage horse.

Your personal profile, when you left university, probably looked more like the Incorporated Member, but has since broadened out as you gained experience. At the time of the review, it probably veers like a jagged saw from one side of the envelopes to the other, dependent on your particular experiences, neither truly one nor the other:

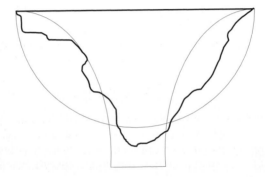

What is vital is that you tailor your preparation and submission to suit the profile you have chosen to demonstrate. The differences are not quite as straightforward as my cartoon profiles suggest, but the figures do give an indication of the fundamental manner in which apparently similar attributes are assessed by the Reviewers, dependent upon which review is being attempted.

As recent changes take effect, I believe that progression from graduate to Member will become the norm, and the majority will gain professional qualification through the Member Professional Review. Once opportunities become available, some will broaden out to exert a significant influence beyond a strictly engineering or technical context and will become Chartered Members through the progressive route.

The timing of these opportunities will depend to a considerable extent on the expectations and needs of your employer and on your developing capabilities as an individual. Some engineers will have been recruited specifically to fast-track into roles represented by the criteria for CEng; others will gradually move into it, while some will prefer to become experts in a technological field. One is not better than the other, just different. One problem seems to be that some salary scales still do not accept these differences or recognise the value of having skilled individuals forming a suitably mixed team.

Not until you are personally satisfied that you fully comply with one of the descriptions outlined in Appendices A should you proceed any further.

Until you honestly believe that you *are* a professional of the appropriate grade, do not apply!

I cannot to see how you can possibly convince others of your capabilities unless you truly believe you are capable yourself! The preparation, drafting and collation of a submission is a long and laborious process, not to be undertaken lightly and surely not without a reasonable chance of success? The idea that you should make an application merely because you are 'time-served,' or seek promotion dependent upon a professional qualification, must not be allowed to cloud your judgement.

Chapter 10

Starting the submission

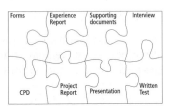

Forms	Experience Report	Supporting documents	Interview
CPD	Project Report	Presentation	Written Test

Once you have decided that you are a professional engineer, capable of becoming an Institution Member, the next question you need to ask yourself is;

'How am I going to prove my competence?'

From this point on, everything you do, everything you write, every form you complete, all the supporting documents you collect, are all aimed at one definite target – proving that you are a professional engineer of the appropriate grade. You must ensure that your jigsaw picture is complete, with each piece locked into the others and with no gaps or overlaps anywhere.

One of the things which ought to have intrigued potential candidates is:

'Why does it appear to be so complicated?'

Why, for example, can you not just submit an expanded c.v. and attend an interview? The answer is that, having interviewed candidates since 1897 and after some 60 years of experience of the review in a format recognisable today, the Institution thinks that this is the best format (at least so far) for candidates to be able to best demonstrate their full abilities. In other words, the format has been refined through long experience to give you the best possible chance of demonstrating your capability as a professional.

So every part matters and must be of use to the Reviewers; what you have to do is find out what the Reviewers need from each part and then make sure you fulfil their expectations.

The first and perhaps rather obvious point about your preparation is:

Make sure you comply with the current rules.

You need to make sure that your copy of the specification or rules is the most recent version. Once you have made the decision to proceed, contact the Institution and check; seek an up-to-date set of forms and documents for the review you intend to take. After all, you would not commence a civil engineering contract without the latest set of documents and standards, would you? And yet I have come across people whose preparation of their submission has been based almost entirely upon hearsay!

Then spend some time reading the documents thoroughly. It never ceases to surprise me just how many candidates reveal during a discussion that they are unfamiliar with detailed points in the ICE publications. It pays to make notes or annotate the document to make sure you are entirely clear which parts relate to you and which are irrelevant (you could do the same with the guidance in this book – but in pencil, in case someone else on a different route wishes to use it!). And then I suggest you read *and digest* the parts you have highlighted as relevant, to be absolutely certain that you fully understand what is required. If in doubt, make use of the ICE Regional Support Team!

Only when you are absolutely clear on how to set about the submission (and, of course, that you are prepared for review) should you start. One important part of the submission which usually gets left until too late is the bureaucracy – the forms.

Filling in the forms

A key component of the submission is the various pieces of bureaucracy. These too are, perhaps unexpectedly, a crucial part of the jigsaw. It is amazing just how many candidates fill them in incorrectly. They seem to be left to the last minute and are completed hurriedly, without any thought being given to why all this information is needed. I presume this is because they are not seen as a particularly important part of the submission. Yet they can help significantly with the overall impression.

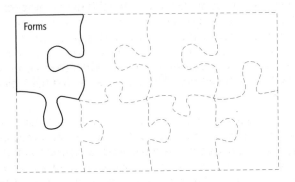

These forms *will* create an impression. Like all the other pieces, this piece of the jigsaw must be used to support your claim to be a professional engineer. Untidy or difficult to read submissions, boxes poorly filled in with an inappropriate pencil (even if the information itself is correct), all conspire to raise initial doubts about your professional capabilities. And certainly do not leave this task to the last minute and hurriedly complete them, only to have your submission returned as non-compliant!

Table A asks you to state your area of technical expertise by ticking one of a series of options. This grid is a match for that completed by the Reviewers, where they indicate to the Institution what they feel capable of examining. This is the way the computer ensures that at least one of your Reviewers is reasonably familiar with your sphere of work. So tick the boxes which best indicate the areas you would like one of your Reviewers to be familiar with.

Table B asks you to indicate your type of employment. The categories are wide and you may feel you do not exactly fit any of them. It does not matter too much; just choose the one which best indicates your employment. This is simply another device to help the computer and Institution staff to choose suitable Reviewers.

The précis

As a manual check on this matching process, you are also asked to submit 'a précis of the significant work section of your proposed report'. I would strongly suggest that this should never exceed one A4 page and in many cases will be only one or two sentences. Why is it needed? It is the means by which staff who know the Reviewers can check that at least one of the chosen pair is sufficiently familiar with your work to be able to comprehend the principles. You need to indicate in broad terms the kind of work that the Project Report covers.

So the précis may only be a short paragraph, if that is all that is needed to give an indication of the type of work covered. Again, once you know the purpose, what needs to be done is more obvious. I have heard of people taking hours to write a comprehensive synopsis because they mistakenly presumed it was similar to that needed for a thesis.

Choosing your sponsors

Your sponsors should be chosen with care, not just a collection of available people from your office. They do have to answer some quite searching questions which you should read before asking people to commit themselves. It does you no favours for the Institution to be told, 'I am not in a position to answer this question' or 'Others may be better able to answer this question'.

One sponsor should normally be the person who trained you, your SCE if you trained under Agreement, but if that Agreement ended some time ago and you lost touch with them shortly after they registered your Completion Certificate, are they in the best position to vouch for you now *as a professional engineer*? You may need to update them on what you have done since they were in close contact, perhaps by giving them a copy of your draft reports and visiting them for a discussion?

While not a requirement, I think it appropriate to ask your sponsors to initial alongside those parts of your experience of which they have direct knowledge – first, it authenticates that experience and, second, it demonstrates that they have seen and read that report and should have decided whether they think you have a reasonable chance of success, something your Lead Sponsor is required to have done.

Do not necessarily look for sponsors solely within your own organisation. Why not ask someone from your consultant/contractor/client/promoter to signify their respect for you as a professional engineer? While you may have had contractual differences and disagreements, surely the Reviewers will be favourably impressed by their opinion of you as a person? Think about the impression you are making all the time – the *purpose*, and choose your sponsors accordingly.

There is now no room on the form for additional sponsors, so it is vital that you choose the best available to you within the criteria set down by ICE. This requires that your Lead Sponsor must be a Member of at least the same grade as that for which you are submitting. The rest need not be Members, but must be members of other Institutions governed by the Engineering Council or with whom ICE has a mutual agreement on qualification.

The Lead Sponsor

You have to nominate one of your sponsors as the Lead Sponsor; it should intrigue you as to why. The reason is that the Institution has become irritated by people sponsoring candidates who are totally unsuitable, often without even seeing the submission, and wishes to be able to reprimand those who do. Sometimes this situation results from an inaccurate or outdated perception of what the Institution requires of candidates; by putting the onus very much on the Lead Sponsors, the Institution expects them to familiarise themselves with the current requirements.

Your Lead Sponsor therefore has serious professional responsibilities. To be properly accountable to the Institution for their opinion of you, they must therefore be a Member of an appropriate grade. Their responsibilities are threefold:

(a) to satisfy themselves that you have a realistic chance of proving you have become a professional engineer;
(b) to certify that your submission is a reasonable and honest reflection of your experience and capabilities;

(c) to check that none of your other sponsors has misgivings either.

So don't put them in an uncomfortable position, give them everything they need to wholeheartedly support your application.

When should the reports be ready?

One implication behind the responsibilities outlined above is that before your sponsors (and particularly your Lead Sponsor) can complete their forms, you must have your submission documents pretty well finalised. Bearing in mind the logistics of asking them if they are prepared to sponsor you, sending out the forms, completion and return to the Lead Sponsor, who puts them in a sealed envelope to go with your submission, the date for completion of this aspect should be about one month before submission. In the UK, this window tends to coincide with family holidays, so do allow plenty of leeway in your programme.

Don't delay! I despair of the number of candidates who apply to sit their Review in the provincial centre last in the list to give themselves another month for preparation; it is mistaken and foolhardy, as technology is enabling the Institution to tighten up the whole process, reducing waiting times. It is hardly surprising that the names of such people do tend to appear in the 'unsuccessful' results list!

If you kept a number of photographs and other material such as calculations and brochures with your records during training, go through them in detail to see which of them, if any, will help the Reviewers – remember the over-riding objective. If they are more of an aid to your recall, then remove them, but perhaps consider taking them with you on the day in case questions are asked which would be better answered with supporting information. You should, in any case, take the originals of any reports with you to the interview just in case there are any queries.

Chapter 11

Continuing professional development

Continuing Professional Development (CPD) is defined by the Institution as:

> 'the systematic maintenance, improvement and broadening of knowledge and skills, and the development of personal qualities necessary for the execution of professional and technical duties throughout working life'.

It is obligatory; Rule 5 of the Institution's Rules of Professional Conduct states:

> 'All members shall develop their professional knowledge, skills and competence on a continuing basis and shall give all reasonable assistance to further the education, training and continuing professional development of others'.

Currently, the Council's recommendation is that every Member records

an average of five days of CPD each year

obligatory but not mandatory. This requirement is in accord with nearly all the professional institutions and is likely to increase in time. The Council of the Institution does not yet intend to make CPD mandatory, as this would involve expensive routine monitoring and legal sanctions, but it may become necessary. It is, however, *obligatory*, i.e. it is a professional, rather than a legal requirement.

Do take due note of this obligation to average five days each year. If your experience extends beyond the minimum (and most does) then this requirement takes precedence and the stated minimum becomes inadequate.

CPD – why bother?

The requirement was originally viewed by many employers with alarm, particularly at the cost implications of 'sending everyone on courses' – not merely the cost of the course, but the productive time lost and expenses involved. There was initial resistance to what was seen as an imposition. Most have now realised that this was a mistaken overreaction. In effect, everything you now know, which you did not know when you graduated must, by definition, be *continuing* professional development. The problem is that you probably have not had an adequately documented quality assurance system.

Although the Institution has not made it a mandatory requirement, evidence of CPD is required when you join, when you change level of membership, notably from Member to Fellow, and for certain professional functions such as adjudication, arbitration and reservoir inspection. CPD records are also checked for all potential Reviewers.

The Institution may consider carrying out an annual check on a random percentage of its membership. This becomes a realistic possibility once the majority of us record our CPD on-line.

Perhaps more significantly, think about the possible implications if you:

(*a*) give evidence in a legal/contractual case;
(*b*) need Professional Indemnity insurance;
(*c*) are accused of professional negligence (a possibility more likely than hitherto given today's litigious attitudes);
(*d*) are required by a client to disclose the qualifications of sta you intend to use on a project as part of a tender proposal.

Would it not be reasonable for the lawyers, insurers or client to ask what has been done to maintain or enhance professional competence since qualification? How would your firm react if they were asked point blank, 'What right has this employee to be making such decisions?' and they were unable to *prove* (to the satisfaction of the risk assessor) your up-to-date knowledge and competence. Record what you are doing to keep yourself reasonably up to date and improve your competence and you won't be caught out!

What does CPD entail?

Since a CPD record became a requirement, engineers have found that they are in fact already doing at least five days per year, and in most cases, considerably more. I am not the most efficient of administrators, but I manage to remember to record an average of over 20 days a year, very little of which is by attendance on formal courses. Today's rapid changes, not only in technology and the legal framework, but also constant reorganisation and redeployment, mean that anyone who does not keep their knowledge up to date or learn new skills will soon be left far behind. Flexibility and adaptability are vital commodities in today's uncertain market-place; for these, CPD is a prerequisite.

The list inside the front cover of the Institution's record booklet covers all the methods of keeping up to date; they go far beyond 'going on courses'! It is possible to fulfil the minimum requirement without ever attending a course, although there are some things which are best covered in that structured environment. The only proviso is that, since the system is self-certifying, it is preferable for there to be some tangible end result, to use as proof for audit if ever needed.

In practice, therefore, the requirement is not an additional imposition, simply that you properly record what is probably already being done. Many organisations now produce quality assured projects, using quality assured products. Yet few staff are themselves properly quality assured. But isn't this the next logical step? Several organisations have already incorporated Training Agreements into their QA systems; all other staff must surely follow.

What CPD is needed?

CPD is a method of continual self-development; it should be an intrinsic part of your continuing training and development, not something which is a 'bolt on' extra. It is necessary to routinely examine your current levels of skills and knowledge and measure these against those that already are, or probably will be required in the future. This is not difficult if done in a rational and formal way. Many organisations now have formal systems for annual assessment of their staff and their training needs, but if yours does not, do it yourself in the Institution's Development Action Plan. You then need to record what you actually did to make good any identified deficiencies in your Professional Development Record.

If your employer already has a documented system in place for routinely (perhaps annually) assessing the effectiveness of last year's training and planning next year's, and is willing to allow it to be used by you for the review, then don't waste time duplicating information. Get permission to photocopy their record. But beware! Some companies only record formal CPD such as courses and seminars, partly because they are entitled to seek reimbursement should you leave shortly afterwards. Such a record will only show a small part of the total effort you are putting into keeping up to date and developing.

The Institution has produced a self-assessment document to assist in identifying areas of your capability which need attention – *Management Development in the Construction Industry* available at modest cost from the Thomas Telford Bookshop in the Institution building. It looks bewildering at first but I have, from personal experience, found it very easy to use and it certainly identified my needs! Candidates are increasingly including the detachable assessment sheets from this booklet as part of their written submission. It seems a good idea to *demonstrate* clearly and very concisely how you identified the needs and what you did about them.

Key criteria are continuing, relevant and progressive.

Going on a basic site safety course three years after first working on site is a sinecure – merely 'complying with the rules', neither progressive nor relevant and therefore probably unacceptable. While it will not be a reason for failure, it could well be held in evidence against you.

A strong indicator of relevance for one critical aspect is supplied by the Development Objectives. How are you going to develop the required competences? Some things are best learnt by 'hands-on' experience, others are best learnt about on courses, and then put into practice. I personally would expect most candidates to have had formal training in the fundamentally important matters of hazard and risk identification and management, and probably the commercial context of their work. These, and similar topics are too important to be learnt solely by experience – they could be expensive lessons!

The requirement for continuing and progressive education and training was probably the origin of the (actually non-existent) 'rule' which I still hear mentioned: that ten days should be spent on technical training (the early career), ten days on managerial (the latter stages of formation), and the remainder should be a mixture of both. Since most young engineers start by utilising the technical knowledge gained from their academic course and gradually move towards a more managerial role, this is probably what will happen anyway, but it is not a requirement.

Also look carefully at the personal qualities being developed, which are necessary to many other professions; things like time management, team working, personal relationships, negotiating skills, running a small business, a foreign language. You do not necessarily need a civil engineering course to learn the basics of these. Consider material and courses beyond civil engineering – the local Chamber of Commerce, Further and Higher Educational courses (both vocational and recreational), distance learning material from other institutions.

What the Reviewers will be looking for is a concerted and considered programme to put right areas of your expertise which you have identified as needing attention. If you find using the English language difficult (and many engineers do) then have you tried to improve by attendance on a suitable course? If you have had difficulty with contractual disagreements or public meetings, have you been on a Negotiating Skills or Dealing with the Public course? If you went to work in a laboratory, did you first find out what the hazards were likely to be and how to avoid them?

Most employers put new employees through an induction process, to explain the workplace and what is expected of them. This is, of course, by definition professional development – have you recorded the lessons you learnt?

Do keep in mind the Council's current recommendation – an average of five days per year. This may prove to require more days than the specified minimum if, as is often the case, you have taken rather longer to reach that key point in your professional development. Again, of itself this would not be a reason for failure, but it is going to raise doubts about your commitment.

CPD is another and important part of the total jigsaw, not an appendage! It is part of your overall development and will be consistent with your reports.

Chapter 12

The Project Report

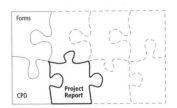

Demonstrating capability

Having watched well over 2000 candidates prepare their submissions, I have concluded that, contrary to how most candidates plan, this is the best report with which to commence, because it is the easier one to write.

This report is to enable you to demonstrate that, as a result of your experience, you have developed, and could use, all the capabilities needed by a qualified member of the Institution of the grade for which you are applying. If the Project Report demonstrates every attribute which the Reviewers are hoping to confirm, then it becomes easier to identify and profile the particular experiences which led to that competence in the Experience Report.

Choose the project as a 'carrier' upon which the attributes can be demonstrated. Work backwards from what the Reviewers are seeking to which of your most rewarding project roles would enable you to best demonstrate your capabilities. Use an abbreviated list of the keywords listed in Chapter 8. This avoids the time-consuming process of writing far too much and then having to edit it down to fit the specified word count.

The name 'Project' Report did give rise to all kinds of misunderstandings and misconceptions. There were residual beliefs among many senior engineers (based perhaps on their submissions of years ago) that:

❑ candidates needed a design, preferably in reinforced concrete;

❑ the report had to include site and design work, which often led to two or more disparate 'projects';

❑ the report had to cover all the aspects of your experience so far;

❑ it had to be a whole project, complete from conception to construction.

These assumptions were just not true! You can use anything which enables you to display your competence. Indeed, there was at one time, talk of calling it the Competence Report. The content might be your entire job, or bits of various recent work, or one extensive project.

The rules say that you should have had a major involvement and some degree of responsibility, such that you are able to demonstrate the required attributes. In other words, this report must *demonstrate* either that you have or that you could (given the chance) readily and confidently take the responsibilities and display all the attributes of a qualified member.

This is an interesting and difficult concept for those well schooled in scholastic examinations, where the emphasis seems very much on the transmission of as much knowledge as possible or on 'getting the correct answer'. Few academic examinations measure potential; most retrospectively measure what you have learnt and can put into practice. At the review, there is an element of potential capability, upon which the Reviewers must make a judgement. You are attempting to answer the question:

> **'Would this candidate, when placed in a position of responsibility, make the correct decisions?'**

Note that the wording is 'would' not 'has'. As I said in Chapter 6, item (c), you are not yet a qualified member, so may not yet be allowed to take decisions commensurate with your sought status. But you must demonstrate that you could! That you know enough about the background of decisions being taken on the work, by those senior to you, to be capable of making similar decisions in future.

I stated categorically at the end of Chapter 9 that you should not attempt the review until you know that you have the capabilities of a qualified member of the grade you seek. The way to determine whether this is the case or not is to work back from what you are trying to demonstrate and choose something which will achieve that aim.

So start from the attributes of a member qualified at your target grade; how can they each be demonstrated through the intended work? Use your Training Record and other accumulated documentation to provide inspiration and information. Make notes! It is surprising how readily ideas can be lost if they are not recorded at the time you think of them. This detailed stage will take time, so don't leave it to the last minute! Indeed, you ought perhaps to have done the exercise in essence before you committed to the review?

The report need not be on one 'job' – it can be based on several from your most recent experience, but be careful in your choice. Remember, you are going to demonstrate *your* abilities and the more words used to describe the work, the fewer remain for this prime purpose. Generally speaking, more than two different projects are difficult to manage in the short presentation, although again, this is not a 'rule' – one candidate successfully delivered seven projects in the 15 minutes!

You will probably write the greater proportion of your report on your most recent experience. I believe you should for three reasons:

(a) you are probably a better engineer today than yesterday – 'old' experience is therefore less likely to demonstrate your full abilities;
(b) you are likely to be more familiar with current work, and hence better able to discuss it in detail at interview;
(c) documentary 'evidence' is probably more easy to assemble, since it is likely to be on your desk.

The work you use need not be 'complete', i.e. from problem to handover, but it must have a clearly defined 'start' and 'end', i.e. 'Preferred route for Exchester Bypass'. In the lengthy progression of a highway scheme from identification of need to construction, the preferred route is a definable stage, so it does have a start and an end.

The work need not be particularly large or 'grand'; the overriding requirement is that it must enable demonstration of your attributes as a potential qualified member – technical or managerial complexity are more important than prestige.

All engineers tend to take what they do for granted; they are essentially problem solvers; once a problem is overcome, they move on to the next one and the original problem slips into their subconscious; only the solution is remembered. It is therefore difficult to recall everything of relevance. Hence the importance of good diaries and training records and the need for time to unravel the original problem from your records and memory.

Once you have completed making notes on the individual attributes, as outlined in Chapter 8, you will have at least sufficient material for this part of your report. The next stage is to organise the information into a coherent whole. Your notes comprise a list of ingredients. It is unlikely that the resulting pudding (the report) will look at all like the ingredients, but the Reviewers should be able to detect all of them as they read through. Tackling it this way will ensure that you write mostly about your role, responsibilities, understanding and experience and a minimum about the project(s), and thus avoid a lot of time-consuming editing.

Minimising the word count

To further minimise the 'waste' of words on explaining the project, make judicious use of sketches, maps and plans or photographs, each one chosen with care to illustrate exactly what you need. If figures (traffic counts, flows, areas, etc.) would be useful for the Reviewers to better appreciate the scale of the work, then annotate the sketch or plan as appropriate. By doing this, you retain as many words as possible to demonstrate your influence on the progress of the work.

Too many candidates sprinkle photographs liberally around their reports without any clear purpose beyond the vain hope that they might impress the Reviewers. Beware! Pictures of the job, however prestigious, will certainly not impress.

The ideal is to refer to a definitive picture, perhaps annotated with key facts about the work (possibly on the front cover), then briefly and succinctly describe the project and the background to it in the first half (no more than two-thirds) of a page, so that you can start on your role within the first 100 words or so. This is not always possible, but it is a target.

I saw the ideal scenario recently, where a candidate was using a demolition project. The front cover of his Project Report showed a photograph of a tower block from the roof of an adjacent block; the proximity of other blocks and the playground, car park and single-storey shops around were all too apparent; even the number of floors could be counted. The report started 'My job was to demolish this block without undue disruption'. No mention of how many floors it had, what it was made of or what the problems were; they were all evident in the carefully chosen picture. The back cover was a pile of rubble, with the perimeter fence, shops, car park and playground still intact! Very few candidates have this sort of luck, but it is something at which to aim.

Much more about the supporting documentation is included in Chapter 15.

Chapter 13

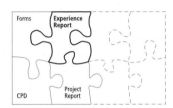

The Experience Report

Again, your curiosity should have caused you to wonder if, in your Project Report, you have demonstrated to the Reviewers that you have acquired all the attributes you need to be successful, why do they also require an Experience Report?

This report is to demonstrate how the attributes were developed through your experience.

Remember that it is relevant experience you are demonstrating; this may come from life beyond the workplace, i.e.

❑ gap years,

❑ work in other environments,

❑ leisure activities, etc.

This report is rather more difficult to compile than the Project Report, primarily because you have too much information: routine training reports over three or more years, CPD reports and, hopefully, cuttings and abstracts from any relevant articles. It is equally important that you address the fundamental obstructions outlined in Chapter 6 before going any further.

Like all the other pieces of the jigsaw, the report must not be written in isolation: you will be dependent on any records you have, one application form requires a c.v., you have drafted your Project Report and you need to consider what supporting documents you might need. It will also tie in with your Continuing Professional Development record (Chapter 11). Don't be one of those candidates who demonstrate a slapdash or haphazard approach to the review by having inconsistencies in the detail of their submissions.

By now, you will have had considerable experience, probably something in excess of five or six years, covering a significant number of projects. It will not be easy to record all this experience in one relatively short report. From personal experience, the Institution will not allow any dispensation, however many years' experience you are trying to describe. Yet everything must be included; you cannot discount experience, every piece of which makes us what we are.

Most of us have difficulty in writing succinct yet comprehensive reports, partly because there is a growing, and hugely mistaken, tendency to believe that communication requires you to tell everyone everything immediately. How on earth did the historic Empire builders manage without fax, e-mail and the ubiquitous mobile phone? The answer is that they knew how and how much to communicate and how to delegate and develop trust. Our society today is bombarded with verbosity on all sides, particularly from political spin in the news and instant fame on television and radio. All of this new technology has become obstructive to proper communication, because it is used badly.

So we all tend to fall into the trap of using a paragraph where a sentence would do. This example (written by a Chartered Engineer) arrived on my desk shortly before I left a previous role; I have no doubt that, like me, everybody could cite many similar cases:

Highway Improvement Schemes

Reference my letter of late November, on the above subject, due to pressures brought to bear the situation has been eased and we can now go ahead with repayment works in 19??/?? for which orders have not yet been received.

I would however remind you, if indeed that is necessary in the present climate, that because of stringent financial limits we are operating under our direct labour resources are consequently severely restricted. Repayment works in general involve us in much greater expenditure of our direct labour due to cabling and jointing operations, than the provision of ducts and jointing chambers, etc., the latter being done by contract under our supervision. For this reason and because of difficulty in acquiring the necessary and in many cases special stores at short notice may I ask for your continued cooperation in taking due note of the time factors we state when furnishing you with estimates and quotes for projected schemes.

Yours faithfully

[157 words]

This actually contains one piece of information and one request; everything else is irrelevant *for the reader*. So a much shorter note could transmit:

❏ the restart of repayment works, and

❏ the request for cooperation

without obscuring either by irrelevant detail. Using his words, the note should have stated:

I am pleased that the problems outlined in my letter of late November have been resolved; we are now able to undertake orders for repayment works in 19??.

Your continued cooperation is sought in taking due note of the time factors in our estimates; this will minimise costs and avoid possible delays in materials supply.

[55 words or 35%]

One-third of the length! And it could be shortened even further by a better choice of words.

Having resolved the problems outlined in my letter of late November, we are now able to fulfil orders for 19??.

To minimise costs and avoid delays, please continue to respect the time factors in our estimates.

[36 words or 23%]

So keep a constant watch on irrelevant detail. Keep asking yourself:

❏ Do the Reviewers need to know this?

❏ Is this information necessary as background to my competence?

❏ Does each sentence make a positive contribution to the overall objective?

❏ Does each sentence use the least possible number of words?

The production of this report requires you to be absolutely clear about exactly what it is that you are trying to prove so that you do not waste any of the words. It will take time, help from colleagues and friends and a great deal of editing. Quizzing many candidates, after their reviews, about what advice they would give to the next group, the answer is almost invariably, 'It takes twice as long as you expect!' In my opinion approximately four months is a reasonable and realistic period, although I well remember one colleague who successfully wrote both his reports in a fortnight!

Purpose of the Experience Report

The guidance suggests a word limit which is tight, requiring you to focus very precisely on what you are required to demonstrate. The purpose is to answer two fundamental questions:

(*a*) Have you had adequate opportunities for relevant experience?
(*b*) Have you benefited adequately from that experience?

Consider each of the answers to these questions in some detail.

Adequate experience?

This first question is relatively easy to answer, since it is factual, based on a list of your appointments and the projects you worked on. But beware! A c.v. alone is not enough! What you must spell out are your precise responsibilities; for example: 'I was the sole Engineer's representative (ARE) on site' is a statement of fact, but does not indicate the precise level of your personal responsibility. What responsibilities were formally delegated to you and how was everybody (including you) informed? Did the Engineer (or their Representative) come out every day to see how you were getting on? Did they come out once a week, or in contrast, did you go to the Engineer whenever you felt it necessary? All three situations represent completely different (and increasing) levels of responsibility, but any one of them is implied by the initial statement and so there is immediate doubt and uncertainty about your exact role.

'Setting out' is another common and bland statement which masks a multitude of responsibilities. You were probably personally responsible not only for making sure that the work being undertaken was in the correct position and of the correct size, but may also have had considerable responsibility for quality control of, for example, component assembly, material quality or the condition of samples.

The limit on words does not allow you much scope to answer both of my questions, particularly if your experience covers several years and many different projects. There is a good deal of purely factual information to be transmitted.

To help in overcoming this problem of limited space, I suggest you write a tabular introduction (which is therefore not included in the word count) to detail much of the factual information which will answer the first of my questions. This approach will release many words in the report for use in answering the second question. The format which seems to work best is a one-page, three-column 'Foreword'. There used to be a requirement for this foreword, but now you are left to make your own decisions on how best to transmit the necessary information.

The best format I have seen looks like this:

Dates from – to –	Projects worked on –	Personal responsibilities
Employing organisation	size, throughput, value	(not 'tasks')
Job title	('clump' if necessary)	
Line manager	Place names not very	
(professional	helpful	
qualifications)		
Line manager's job title		

The tabular format enables you to reduce factual descriptions to gain space for the benefits you derived. It gives the Reviewers an immediate

❏ overview of the sort of professional environment you were working in,

❏ indication of the range and scale of the projects you worked on,

❏ precise details of your actual responsibilities.

Some points to bear in mind if you decide to complete such a table are listed below.

In general   Never use words like 'several' or 'majority of', which do not give the Reviewers any indication of size. Give a figure! Even if it is not precisely correct (and how are the Reviewers to know?), '5' oers the Reviewers a realistic idea of the total, which is of a very dierent order of magnitude to '15' or '50'!

First column   Your job title may mean something within your organisation, but does not necessarily give any indication of your responsibilities to anyone outside it. 'Principal Engineer', for example, can mean very dierent things in dierent companies and authorities.

Include your line manager to give the Reviewers an indication of the professional environment in which you were operating. If your line manager was a member of ICE, then (M) or (F) will suffice, but if they belong to other Institutions, then their title must be given in full – MIStructE or RICS.

Second column   'Value' may not necessarily be cost, but could, for example, be the quality clean-up of a stated quantity of wastewater (e.g. 5 m^3/s) or the number of vehicles diverted around a town. If you oer a project cost, then do also give the proportion which was the cost of your part of it. For example, as much as two-thirds of the total construction cost of a wastewater treatment works could be the mechanical and electrical equipment.

'Trumpton bypass' might mean something to a Reviewer who has been there, but '$4\frac{1}{2}$ mile rural dual carriageway' is significant to anyone involved in roads and is very different to, for example, a '$4\frac{1}{2}$ mile urban dual carriageway'.

Third column   By far the most dicult column to complete is the third – your responsibilities. It is all too easy to write things like 'design' or 'setting-out'. These are not *per se* responsibilities, but the means by which you discharge your responsibilities. Setting-out is the means by which you ensure 'positional control and accuracy' and design is the process by which you achieve 'fitness and acceptability for a predetermined purpose and acceptable risks'.

To help to explain the above advice, here is an example of a submitted report from a candidate who was successful. (As a matter of principle, I do not use any examples from unsuccessful candidates; they have suffered enough. I have also removed any references which may identify the guilty!)

On commencing employment with Midshire County Council Highways Department, Design Services (Roads) Division, I was assigned to the Major Improvements Section and gained much experience working on several highway improvement schemes including:

(*a*) designing accommodation works for the Bugwash Bypass and compiling the accommodation works bill of quantities using the MICRORATE software;

(*b*) preparing vertical alignments for various bypass schemes;

(*c*) producing working drawings, schedules and contract documents for a section of the Feetwet Ferry Southern Primary Route Improvement Scheme – the working drawings were drafted on the Computer Aided Design system;

(*d*) designing preliminary horizontal and vertical alignments and junction layouts, using Department of Transport design standards and producing approximate cost estimates for a proposed bypass to Middlewallop Town Centre;

(*e*) preparing cost comparisons for two alternative junction improvement options.

For the majority of these schemes, once I had been briefed by the Project Engineer, I was responsible for obtaining relevant data from various sources, performing design and compiling drawings and documents.

Most of this information could, with strict editing, be incorporated into a foreword table, leaving most of the space to enlarge upon the two unsubstantiated statements 'gained much experience' (first paragraph) and 'I was responsible for obtaining relevant data from various sources, etc.' (last paragraph), which, as they stand, are practically useless!

Sept. 97 to Apr. 98	Bugwash rural bypass (£4M, accommodation works £150K)	Confirming and detailing accommodation works including quantities (MICRORATE)
Midshire County Council		
Highway Design Services	Flatland rural bypass (£6M)	Choosing vertical alignments for this and three other proposals
Major road improvements	Feetwet Ferry Southern Primary Route (£12.5M)	Production of detailed tender documents using CAD
Graduate Engineer		
Fred Bloggs, Principal (M)	Middlewallop town centre bypass (estimate £13.5M)	Preliminary vertical and horizontal alignments and junction layouts
		Cost comparisons for two alternative junction layouts using previous rates

This foreword actually tells the Reviewers more facts, while removing 131, or 80 per cent, of the words from the extract, which may then be replaced with an indication of the benefits gained from this experience (rather than the candidate's original wording, which actually tells the Reviewers very little).

The foreword gives the Reviewers a clear skeleton of your career through which they can more readily assimilate the contents of your report. Do try to keep to one page! Anything more is becoming an addition to the report and should be an appendix, which is nowhere near so effective.

Demonstrating the benefit

A foreword table enables you to omit much of the explanatory text in the body of the report. But don't fall into the trap of wasting this valuable space by repeating the information again in the text. Just a few words will link the foreword to the report: 'At Feetwet' or 'On Bugwash' is all that is required. If the Reviewers need to check when, where and what you were, they will look at the foreword. This technique enables you to use the maximum number of words to concentrate on the benefit you gained from the experience. Thus the Midshire candidate could have expanded on any or all of the following:

❏ How was 'relevant data' decided upon?

❏ Where was it sought and found ('various sources')?

❏ How did he choose and learn to use the design software (Development Action Plan and Personal Development Record)?

❏ How did he make the problems fit the software and was any check made that the assumptions inherent in the programme were acceptable?

❏ What form of contract was being used? Why?

❏ How were estimate costs established? Was the base data adjusted in any way? If previous rates were used, did he add a percentage for preliminaries? If so, what?

❏ How accurate did the designs and estimates need to be for realistic comparisons to be made?

❏ What did 'briefed by the Project Engineer' actually mean? There is a huge dierence in responsibility between being told exactly what to do and where to find the standards, and being given a quick overview and left to get on with it!

❏ How did he determine what was 'relevant' data and what were the 'various sources' and how did he find them?

❏ What did 'performing design' and 'compiling drawings and documents' actually entail?

Since the report must demonstrate that you have become a professional engineer, it is important that you concentrate on those parts of your experience where you gained maximum advantage. Not all of your work was useful for your development; some, almost inevitably, may have been repetitive or undemanding. Covered in your foreword, it does not need to be mentioned again in the body of the report.

Much of the benefit you have gained from experience will not have come from your direct involvement but from your observation and questioning of things going on around you. This is what the Reviewers need to be told about in the report – not what you were doing (they already know most of that from the foreword) but what you were learning.

The major value of your experience may perhaps be towards the latter part of your career to date, when you

❏ realised what becoming a professional engineer entailed, and were therefore better able to derive maximum benefit;

❏ were probably carrying greater responsibilities, more representative of a professional engineer.

If your mentor or SCE is familiar with your work, it is all too easy to 'read between the lines', because they know what was involved, rather than what has actually been written, which tells the Reviewers very little. 'Once I had been briefed by' is a typical example of this; both trainee and SCE know what this briefing involved, but the Reviewers do not.

> **The 'but it's obvious' syndrome must be overcome – competence must be *demonstrated*!**

Supporting documents

One other way of saving words is to use diagrams and sketches; this avoids lengthy explanations of site layouts, locations or details within the text. I give guidance on how to assemble this part of the jigsaw in Chapter 15.

Chapter 14

Reports – common faults

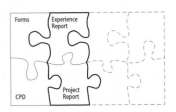

Underplaying your hand

Nearly every submission report I have read underplays the writer's hand; there seems to be a reluctance to spell out exactly what you did and understood, expecting the Reviewers to 'read between the lines' to decide for themselves. This is not good enough – you must *demonstrate* your abilities, not assume that they can be tacitly inferred. It is not generally in an engineer's character to boast, but in this instance, you must try! It is also difficult to admit to mistakes, yet these are often where good experience was gained. The Reviewers will not be so interested in the mistake as in what you did to rectify it and what you learned from it.

It took me some time to unravel why most candidates underplay their hand, but I think I know one fundamental reason. Engineers are problem solvers; once a problem is resolved and a solution found, they go on either to solve the problems of implementation or to another problem. In other words, they forget the original problem. So to write the report successfully requires you to remember the solution and then unravel the thinking behind those decisions to fully expose the original problem.

The difficulty becomes greater the more experience you have, because you may then know how to solve a recurring problem, i.e. it is not a problem to you any more. But it still remains notionally a problem, enabling you to demonstrate how your experience enabled you to solve it!

Another difficulty is that we all know that, in reality, none of us works in isolation. Everything we do, we discuss with others. As a result, we are loathe to take personal credit for our work. But ask yourself who would have taken the blame if things had gone wrong – if it was you, then you were personally responsible, however many people you discussed the problem with!

Every time you mention that something was done or happened, immediately ask the questions 'Why?' and 'What else was considered and why were the alternatives rejected?' as well as being sure to explain your precise role in the process. Be prepared to put yourself on the line and tell the Reviewers that, with hindsight, you now believe that there might have been a better way (after all, one of the things you must demonstrate is the ability to learn from experience). Even where the decisions were not yours to take, you can state that you offered advice, collected information, thought of ideas or suggestions or made recommendations which were subsequently accepted. You may even have drafted the instruction, letter, report or similar documentation for someone else to sign – say so! And where you had no direct involvement, you must show that you understand how the decision was reached; after all, it will not be long until you are making comparable decisions as a professional engineer. It is always better to have some experience *before* decisions need to be made than afterwards!

An example:

> The bridge was designed as a continuous, twin-celled cast in-situ reinforced concrete deck supported on piers with piled foundations and two spread footing abutments. The contractor proposed an alternative single-celled design in his tender which the client accepted.

What a throwaway! And perhaps a tacit suggestion that the candidate's original design had not really been thought through? I think the Reviewers are almost bound to ask, 'Was that not something you considered at preliminary design stage?' The candidate must know why the original format was chosen, what alternatives were rejected and whether one of them was a single-cell design. Why was the decision changed when the contractor offered the alternative? The candidate may not even agree with the decisions, but must demonstrate understanding of them.

I am often told by candidates that the reason why such explanations are not included is because they have been advised to leave questions hanging for the Reviewers to ask. I strongly advise that such a deliberate approach is mistaken for two reasons:

(*a*) the prescribed length of the reports is too short to include everything, so there will inevitably be unanswered questions without any deliberate attempt to pose them;

(*b*) the Reviewers are wily enough to recognise a 'trailing coat' when they see one and will probably avoid it, because they know that you know the answer! There will be plenty of other questions which are not 'flagged up' so blatantly.

Another example of a case where the candidate hopelessly underplayed their hand:

> The piled foundations for each of the piers consisted of driven vertical and raked steel H piles founding at varying depths across the valley. In addition to the seven permanent piers, eight temporary piers were erected in between the permanent piers, again supported on steel H piles. Thus I gained much experience monitoring piling operations.

When I discussed this experience with the candidate, I found out that he had in fact, devised a neat and simple device to aid the checking of the required sets on so many piles, thus saving hours of laborious and repetitive work. Yet he made no mention of this in the report! His last sentence was wasted.

Writing in the third person

By far the most common mistakes relate to the manner in which you write. I can almost guarantee that you will repeat the most common of all – you will say, 'I was involved in . . . ' or 'I was responsible for . . . '. To avoid these vague generalisations, *always* turn the sentences around to force you into being more specific – 'My involvement included . . . ' and 'My responsibilities were . . . '.

Other very common understatements include phrases such as 'I was then transferred to . . . ' or 'Once I had been given the task of . . . '. How much more positive it sounds and how much more in control you appear to be, if these are rephrased as 'I then transferred to . . . ', 'I had the task of . . . ' or even better 'I did . . . '. And the changes actually *save* words!

Many engineers have written reports for committees or clients, or for an academic forum, where writing in the third person is invariably required. There is frequently a tendency to write these submission reports in a similar manner as a result of such experience. The worst scenario is, like this example, where the report actually gives the impression of being copied from a textbook:

> Control of a project is a continuous process which may be represented as shown. As a job proceeds performance should be monitored against the targets defined by the plan. Monitoring itself does not constitute control; this is exercised by making decisions on actual performance and updating the plan.

But the most usual third person style is exemplified by:

> Many questions became apparent to me. How long would the design take? What resources would be required? What information was needed and at what stage?

Remember the underlying purpose! Rewrite as:

> I identified what resources could be available and decided on a realistic time-scale within which I could programme the receipt of key information.

The length of the sentence has been decreased but, much more importantly, your responsibility is now *demonstrated.*

Use of abbreviations

Use acronyms whenever you can to keep the report short, but always tell the reader what the letters mean the first time they are mentioned. In a previous example on page 76, what exactly is an ARE? You might expect everyone to know, but you cannot assume they do! There is another important point to be made here; since ARE is not a formal description of any person under any form of contract, the Reviewers have no idea of the responsibilities or role the candidate is playing unless you tell them!

Be careful that your acronyms are correct. I hate to think of the number of times I have seen 'r.c. concrete' referred to – I am still trying to find out what it is!

Jargon

Avoid jargon! I know we all use it every day as part of our communication at work, but in formal documentary communication it is unacceptable. Here is a classic example, where I do not think there is one genuine engineering term in the whole sentence:

> The rebar cage was prefabbed outside the hole and craned in just before the RE's inspection so that the rubbish could be removed easily from the shutters.

Such use of jargon is completely unacceptable. But worse – there is an implication here that perhaps the cleaning would not have been done at all but for the RE's inspection. So it is absolutely vital that you read and comprehend exactly what your sentences say, not what you think or believe they say! This version is much better, and uses fewer, more appropriate words:

> The reinforcement was prefabricated alongside the excavation and lifted in after the formwork had been thoroughly cleaned and prepared.

There is no need to mention the RE; the cleaning would have been done anyway, because you must demonstrate that you have a professional attitude!

The final stages

Even when you have edited the drafts to something like the correct length and made sure that every sentence is aimed at the objective of proving you are a professional engineer, the task is by no means complete. Now you have to start checking the spelling, punctuation and syntax. If you have difficulties with these, do not tempt fate by being too adventurous. Like Winston Churchill, who admitted to early difficulty with the English language, keep your sentences short and your words simple, only use full stops and commas. I note that the current (2005) Education Secretary has recommended that schools stop teaching the use of the possessive apostrophe ('s), 'because it is too difficult'! Even though the Institution has advised the Reviewers that they should not penalise a candidate for failures in the education system, I think this is going too far, and would strongly advise that you master the use of 's, or just avoid having to use it.

It is very difficult to read your own work and find errors, because you tend to read what you want it to say, rather than what it actually says. You also know what you are trying to say, whereas someone reading it for the first time does not. So you have to try to detach yourself and read the reports as a third person. To help, it is also a good idea to get anyone, engineers who do not know anything about your experience and even people who are not engineers, to read it. They will ask questions which cause you to query whether what you have written expresses exactly what you intended. For this book, one of the most helpful stages in the sequence of production was when the editor (who is not an engineer but does have a marvellous command of English) started to query what I had written; it really did make me think afresh.

Checking spelling is something else which requires great care and time; it needs to be done as a separate exercise, where you do not actually read the report, but only check the spelling, punctuation and syntax. You cannot rely on the word processor's spelling or grammar checking capabilities which are limited, generally Americanised and usually cannot tell the difference between 'right' and 'write', 'their' and 'there', etc. You need to be very painstaking – this example is a classic, which many people who read the report failed to notice:

This meant that both pupils and their parents were forced to cross a very busy dual carnageway.

Do spread your checking beyond the report itself, to the cover sheet and picture captions. It is not sensible to have a cover sheet which announces in two centimetre high letters that your submission is for the

CHARTERED PROFESSIONAL REVUE

or that your project report describes a

LARGE SCALE DEWARTING SYSTEM

both of which appeared at the Institution!

Summary

By now I hope you have realised that the compilation of these reports is not easy. It is my considered view that there is a fundamental need for time in this process and I usually raise eyebrows when I suggest that at least four months is a reasonable minimum. There is no substitute for reflection – setting aside the report for a few days and looking at it again with fresh eyes. You also need to ask as many others as you reasonably can to read your work – and not only other engineers. They will all offer different suggestions, but at least they will have caused you to look at the report in different ways. You can then make your own mind up as to which suggestions you incorporate and which you discard.

Since, in theory, you have infinite time and infinite support and resources to prepare these reports, the Institution expects them to be perfect, just as any professional report emanating from your office should be. Whenever I ask candidates after their review if they have any advice for those candidates following them, their response is invariably 'I wish I had allowed myself more time!'.

Chapter 15

Supporting documents

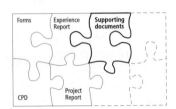

Some mention of the supporting documents has been made already since, as I explained in Chapter 6, it is impossible to separate any one part of the submission from the others; the pieces of the jigsaw must interlock to form a complete picture.

The *Routes to Membership* documents say that 'appendices containing selected examples of the candidate's work should be included to provide evidence of their achievements'. The aim is to provide supporting evidence which

❑ assists in demonstrating the abilities being displayed in your report;

❑ justifies and supports the decisions which you mention in the report.

Careful choice is necessary; the Reviewers are not impressed by sheer bulk. Indeed, one Reviewer deliberately collected statistics to show that

> **the success of a submission is inversely proportional to its weight.**

The inclusion of every document must be a considered decision; does it contribute positively towards the overall objective – to prove you are a professional engineer? If it does not, then discard it.

Analyses

These must demonstrate that your decisions (or the decisions taken by your line management, but to which you contributed) are based on sound knowledge and understanding. They need not necessarily be engineering analyses, but could just as readily be commercial or economic.

Detailed analysis is not design, but is an integral, indeed vital, part of the design process and generally takes place towards the end, when you are justifying the solution you have chosen and making sure it will withstand the forces and loading conditions you have decided it must resist. Prior to that, analysis (generally using quick design methods) helps in the choice of the most appropriate solution.

Documentary evidence is needed for the Reviewers to ratify technical competence if they feel it necessary, but they are much more interested in how the calculations were used as an integral and vital part of solving a problem or implementing a solution. The purpose is to *demonstrate* your understanding of technical (or scientific for Associates) principles, not that you can 'do calculations'!

Do not rewrite calculations specifically for use in the submission; documents should have been prepared during the normal course of your work. Do not abstract aborted calculations; they were an integral part of the design process, vital in choosing the most appropriate solution. Do show that you know how the loading conditions were chosen, even where they are established by others – be able to answer such questions as 'Why was this designed for a 1-in-25-year storm?' or 'Why did you consider earthquake loadings?'

As long as your calculations are neat and legible (even a few crossings out are acceptable), you state where new figures or numbers came from and explain why you did them that way by annotating the calculations appropriately with cross-references, then the Reviewers will be able to see clearly that not only did you do the calculations, but that

you understand what the calculations are doing.

Most calculations are done by computer. The Reviewers do not wish to see reams of printout, but need answers to questions such as:

❏ What assumptions were made or are inherent in the computer program?

❏ How did you satisfy yourself that those assumptions were valid for your problem?

❏ What assumptions did you make in order to ensure that your problem would fit the program?

❏ With hindsight, was the method or program you used the most appropriate?

And, the most important of all:

❏ How did you satisfy yourself that the results were realistic?

These sorts of questions make it quite clear that calculations are a 'means to an end' rather than 'an end in themselves'. You must *demonstrate* how the calculations were used to solve problems such as deciding on alternative structural forms, sizing and reinforcement of members, choice of materials or the most appropriate construction methods and plant. For construction method statements, for example, you will probably have had to consider temporary stability.

In all calculations, assumptions are made about the value of constants and factors. If you, for example, used a value for Young's modulus or a factor of safety, then do be prepared to discuss the fundamental principles of why it was deemed appropriate. It may not have been your decision this time, but in the future, as a qualified professional engineer, you will be required to make similar decisions. You must demonstrate that you can!

Cost data

A conventional cost estimate or bill of quantities might be one 'vehicle' to *demonstrate* an understanding of construction methods and the financial implications of the solution, but only if it is used properly. The compilation itself reveals little or nothing about your competence as a professional engineer, only your methodical administration.

A bill should include something to demonstrate your under-standing of rates and item coverage – the 'build-up'. It is important that you do more than use previous rates from other estimates or quotations so that you demonstrate your understanding of how the rates were adjusted to suit your particular job, or what a rate includes – item coverage. You would not have to be a professional engineer to copy previous rates into a new bill, but I am not at all sure the result would actually be realistic!

The job for which the cost estimate is compiled need not necessarily be a construction project; it could, for example, be the manufacture of apparatus or equipment, or a proposed traffic count.

Other possibilities might include:

❏ the substantiation of a claim;

❏ the estimate for a variation requested by the client or their agent;

❏ the build-up of an estimate for a proposal;

❏ an estimate of design costs.

Illustrations

Use drawings throughout your submission to save words – 'one picture can tell a 1000 words'.

Another example:

> I was temporarily seconded to this site as an ARE. The works at this site were part of several 'Advance Works' contracts for major improvement works to a busy road junction over and adjacent to the London Underground. These particular works consisted of the construction of a pedestrian/cycleway ramp and associated retaining wall from an existing bridge over the underground lines to an anchored sheet pile wall at the entrance to a future subway under a road adjacent to the works.

Can you visualise the site? I happened to know where it was, and even then had difficulty! This situation is crying out for a diagram. Not a copy of the *A–Z* which would present far too much detailed information, but an outline sketch.

And look at the wasted words: surely a secondment can only be temporary? And there are no less than five references to 'the works' in only two overly long sentences – the candidate obviously never read the report aloud! A sketch and drastic editing could release many words for telling the Reviewers how the candidate benefited from the experience at this complicated site.

Large drawings are not a good medium to use for the submission because they

❏ include too much detail;

❏ are extremely dicult to unfold and fold.

Many candidates make the mistake of using contract drawings. I believe it is much better to use simplified extracts from them or the sort of simple diagrams favoured by those who prepare publicity material. After all, contract drawings are required to transmit detailed and accurate contractual information, generally far too detailed for the purpose of your report. In all the many submissions I have seen,

I have yet to find a valid reason for including contract drawings in a submission.

There has invariably been a better option.

Since the Reviewers have to assimilate the information quickly as an adjunct to your experience, they do not want to have to unravel a complicated drawing to glean the basic information they need. To assist them, I suggest that you put any pictures within the text, alongside the part of the report to which they refer; word processing packages make this relatively straightforward.

Other kinds of illustrations might include, for example:

❏ photographs

❏ architect's 'impressions'

❏ extracts from public explanatory leaflets.

Be careful in your choice of the first of these. Photographs have rarely been taken for the purpose of the review (unless you really have been planning well ahead) and so are inevitably second best. Do not be tempted to include a photograph merely because it is the best you have. Keep asking: does it help the Reviewers to understand the situation? I well remember a photograph of a bright blue waterproofed bridge deck, with a faint line across it, entitled 'The expansion joint'. The potential candidate told me, 'It's the only photograph I have of the deck'. I persuaded him that it was useless to the Reviewers, so he substituted a sketched cross-section of the expansion joint!

A normal method of communication for engineers is drawing. So demonstrate that you can visualise an engineering problem (spatial awareness) by including drawings and sketches (not necessarily 'exactly to scale' or drawn with a straight-edge) where you thought out a design, technical or construction problem or transmitted information – e.g. to CAD or to the subcontractor or designer, or to someone manufacturing apparatus.

You *will* be expected to draw during your interview – paper and pencil are provided.

Look very carefully at every illustration. First, to be sure that it does contribute towards the overall objectives, either

❏ demonstrating you are a professional engineer, or

❏ reducing the word count.

Second, look very carefully at the background to every picture: is there anything there which you would rather not show? Even professional journals have frequently been criticised for using pictures where something inappropriate is going on in the background. Too often, photographs open up discussions on peripheral matters like safety or efficiency which can take you unawares.

Passport photograph

You are required to send a passport-sized photograph, with your name written on the reverse, to the Institution as part of your initial application. I suggest it is a good idea to include the same photograph, perhaps scanned on to the cover of your Experience Report so that it does not get mislaid, for the two Reviewers, so that they also know what you look like. I know that many Reviewers appreciate this touch, because after a couple of days spent reviewing six candidates, it is sometimes difficult to remember which candidate was which.

But do have the photograph taken when the review is imminent. Changing the style of your hair and/or its colour or wearing contact lenses on the day when your photograph shows you wearing spectacles can, and has, caused confusion.

Summary

As far as possible, try to incorporate the supporting documents in such a way that the Reviewer can refer to them at the same time as continuing to read the report. Obviously, if the support is a substitute for words, then it should be incorporated into the text whenever possible, but a few pages of appendix giving more detailed justification for a technical or financial decision is better bound in at the back.

One very useful device for overcoming the problem of referring to appendices while reading the report is to print the appendix on the right-hand side of A3 paper (in a similar manner to the ICE Written Test booklet). The information can then be folded out alongside the text and folded back when finished with. This is particularly useful when you are paying for photocopying, since any colour photographs can be pasted into the text afterwards.

Minimise your supporting documents. Do not include repetitive calculations. The Reviewers are not going to check whether you can do arithmetic! They are far more interested in how and why you did the calculations, and what values you chose for any constants, load cases, factors of safety or profit margins.

Chapter 16

Putting the submission together

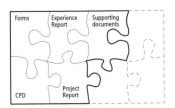

By now you have compiled five of the eight pieces of your jigsaw and the picture should be becoming clear. Now is the time to finally make sure that all these pieces interlock without either gaps or overlaps. Each document should stand alone as a complete entity, but be referenced to the others. Supporting documents should be rechecked to make certain that they really do *support* your case or provide the Reviewers with essential background information. It is at this stage that, as a Regional Liaison Officer (RLO), I used to get frantic phone calls:

'My submission weighs more than 1 kilogram!'.

What these callers have done is put their entire submission on the scales and frightened themselves because they have only scanned, rather than read and understood, the specification.

You are in fact going to send three parcels – one by the 15th (UK) of the submission month to the Institution which will weigh much less than 1 kg and, later, two more to your Reviewers. It is these latter submissions which have a weight limit. In my extensive experience,

properly focused submissions do not approach the limit.

The rules are rather like the Highway Code – you cannot be prosecuted for breaking the code, but behaviour in contravention of the code can be used in evidence against you if you subsequently have an accident! Very few Reviewers to my knowledge, actually check the weight, unless the submission feels heavy (remember, they have to carry several around) or is full of superfluous material. They merely compare it with the other submissions they have received or, being engineers, just check the postage!

Discussing the three parcels leads me to another point: it is a good idea to draw up a table of who needs what documents, so that you can readily see how many copies you need of each. I compiled this one in 2004 for the Member Professional Review. It is up to you to check that the rules have not changed since then.

Document	ICE admin.	Reviewer 1	Reviewer 2	Self	Total
Application	1			1	2
Fee	1				1
Sponsor support	3				3
Academic qualifications proof	1				1
IPD proof	1				1
Précis	1			1	2
Photograph	1	Scanned	Scanned		3
Project Report		1	1	1	3
Experience Report		1	1	1	3
Supporting documents		1	1	1	3
DAP		1	1	1	3
PDR		1	1	1	3

Conflict of interest

You are given about four weeks' notice of your review date. In the same letter, you are given the names and addresses of your two Reviewers. With your lead sponsor, who probably has a wider perspective, check that there can be no possibility of any conflict of interest, such as your respective employers being in serious contractual dispute or perhaps simply that you have met previously. Do not be afraid to bring any conflict of interest to the attention of the ICE Reviews Manager. It is probable that the Reviewer concerned has already done the same anyway!

Covering letter

Get your packages off to the Reviewers as soon after receipt of the letter from the Institution as possible. The minimum requirement is to allow two clear weeks before your interview. But it surely creates a good impression of organisation and control if you get your parcel to your Reviewers sooner, particularly remembering that they are going to read the documents in their own time.

Would your organisation send any documents to a client without a covering letter? Yet many submissions arrive cold and unannounced. Include a letter, addressed personally to each Reviewer (and copied to the other so that they both know they have received it) in which you say that your submission is enclosed in accordance with the rules (quote the paragraph number), draw their attention to any deviations from the rules and give reasons for them, and in the final paragraph, say that you are looking forward to meeting them at the prearranged date and time. This will prompt them to check that their information agrees with yours and removes the possibility of any error.

Make sure that your parcels arrive – and in one piece! Since you cannot phone your Reviewers to check, send the parcels by some form of recorded delivery so that another uncertainty is removed – you will know they have arrived safely. I know that the Institution suggests a reply-paid postcard, but in my experience many Reviewers are not at home when the package arrives, and it may not be until at least the following weekend that they open it. So my advice is to use one of the many tracker systems available. You don't need the stress of the suspense!

Package the submission documents securely in a purpose-made envelope (available at all stationers and the Post Office); this is so much better than constructing a parcel. Reviewers like to use the envelopes as a filing system, keeping each candidate's documents together, so it also helps if you put your name and candidate number on the back flap in felt-tipped pen.

Make absolutely certain that you use the Reviewers' names and qualifications exactly as you were given them. There is nothing more annoying than receiving a letter that is incorrectly addressed or with your name misspelt and, because of its familiarity, it is something which the recipient notices immediately. Take the Reviews Office letter with you, so that if a Reviewer does complain, you can show them precisely what information you were given.

Waiting

It seems an eternity from the final date of submission to your interview. This is the time to look ahead, to try to anticipate how the remaining two pieces might be compiled; you are not totally in control of these but, as far as possible, you must reduce the chances of the unexpected. For example, you should be able, perhaps with help from other people in the office, to make a realistic guess as to what topics might be covered in the written test (if you are having one), since it will be based around your own personal experience.

Researching the Reviewers

It is always a temptation to try to find out who your Reviewers are and what their particular areas of expertise might be. This is, perhaps, an understandable effort to remove yet another uncertainty, but I fear it is misplaced and can, on occasion, backfire. You will not get any information from the Institution. I, and my colleagues in the regions, do get infrequent phone calls asking us about Reviewers; we are always noncommittal in our responses!

The reasoning is straightforward. Your Reviewers have been chosen for their expertise as *Reviewers* and not particularly because they are able, from their own particular field of work, to ask searching technical questions. Remember the purpose of the Review – to explore whether you have become a professional engineer, not whether you can technically do your job.

All the Reviewers are volunteers, who serve the Institution in their own time and largely at their own expense, apart from travelling and accommodation expenses. They do it for two reasons: one is their professional responsibility and a desire to serve their Institution in a tangible way, the other is that they genuinely enjoy meeting 'the next generation' and really look forward to successful reviews. The one thing they do not want is unsuccessful reviews!

Most were trained themselves some years ago and all have an interest in training and development and the maintenance of the Institution's reputation for excellence. The Institution is very proud and protective of its Reviewers, who are themselves subject to continual monitoring and review. Not many are, strictly speaking, volunteers – most are spotted as potential Reviewers by the regional support teams and asked to make the commitment, usually because of their involvement with younger engineers' training. Their initial response is almost invariably a doubtful, 'Do you think I am good enough?'.

All recruits go through quite a lengthy process of training, at any stage of which the Institution may, without prejudice, decide not to continue. All Reviewers are required to undergo formal training at least every three years; most in fact do rather more than this minimum. And many are recruited as young as their thirties – so don't presume your Reviewers are inevitably going to be old.

The difficulties in researching your particular Reviewers come when you perhaps find out that one has a reputation (probably totally false, because you cannot possibly sample a representative group of their candidates) for a particular line of questioning and you become obsessed with that to the detriment of other aspects. The interview will, almost certainly, be perversely different! Or you might hear that a Reviewer is particularly aggressive (again, probably without any real justification) – how is that going to affect your build-up? Hardly likely to give you confidence, is it? Having watched and participated for several years, I know that most Reviewers do in practice change their style slightly to suit each particular candidate.

The final danger of research is that you can easily be misled. One Reviewer has spent his entire career in the South and Midlands. One candidate found this out and felt very secure describing and drawing a problem in the North; he was disconcerted (to say the least) when he realised that the Reviewer knew the area well – because close relations live there!

Of far greater significance is deciding how to make the presentation and what it should cover.

Chapter 17

Preparing for the presentation

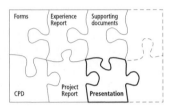

The presentation is the key element locking the submission documents to the interview. It is a great opportunity to influence the whole process; it will occupy about a quarter of the interview and could well determine the shape, content and style of the remainder. It is the first opportunity for you to actually show yourself operating as a professional engineer rather than writing about yourself.

The need for a presentation

Why do the Reviewers need a presentation when they already have so much information from your submission? The formal introduction of a presentation really only regularised an informal approach by the Reviewers which had been apparent for some time – to start by asking, 'Would you like to tell us a little about your project?'. This approach reinforces the view that the interview is *your* chance to strengthen your case and the Reviewers expect you to do most of the talking, with them only directing you to those matters they would like you to cover.

The presentation has been very effective for the Reviewers. They learn much about you, particularly whether you can inspire confidence, i.e. create the impression that you are passionate and committed about your work and really understand the full effects of what you are doing. The Institution is well aware that many of us have to justify our decisions in a variety of public situations – exhibitions, enquiries, committees and the ultimate, in courts of law. So you are expected to show that you could cope in such circumstances. Of course, you will be nervous; we all are! Even the Reviewers! It is how you control those nerves which matters.

This part of the review is meant to replicate the sort of presentation you will be expected to make as a professional engineer when putting proposals to a client. You will have sent them documents beforehand, but would follow these up with a visit to draw their attention to the key issues and why you believe your organisation is the best one to undertake the work. So in this case the purpose of your presentation is to draw the Reviewers' attention to those aspects of your report where you displayed your full ability and convince them that you really are a professional engineer.

It is not a good idea to sit and read your report verbatim, although it is just about possible if you hurry. It was tried in the early days of the presentation – unsuccessfully! You do need to quickly sketch in the problems and then move on to pick out the key points in the work which *demonstrate* your skills and abilities and reinforce them.

Presentation material

You can use almost anything – so long as it is appropriate and makes a positive contribution to the fundamental aim of the review – to demonstrate the full range of your capabilities. Remember, you are going to talk to two people across a small table. I do strongly recommend you to use something for them to look at – anything which will prevent the Reviewers spending the whole 15 minutes staring at you, which would be very disconcerting.

The almost universal aid is an A3 flipchart, with pictures and diagrams on the Reviewers' side and bullet points on the back. I have formed the opinion that A3 is in fact too large, and A4 is more appropriate. There is no doubt however, that a flipchart is a convenient way of keeping yourself on track, both with content and timing.

Most of the presentations I have seen have used far too much material – too many photographs on one display, too many page displays. After all, you are expecting the Reviewers to assimilate the visual information virtually instantly as an adjunct to what you are telling them, so they cannot realistically be expected to take in much. How many should there be? A rule of thumb which I was given was: no more than one page every three minutes, i.e. no more than five in a fifteen-minute presentation (not including the front cover).

I fear that many visual aids are compiled to assist the presenter, rather than the Reviewers. When being coached as a lecturer, I was told, 'If you have to resort to words, never more than four lines, never more than three words to a line'. A maximum of twelve words on a display – not a great deal of scope. But if you look at advertising hoardings, they do not seem to stray far from that advice.

In the review scenario, I would suggest that a movie projector or video with screen are both inappropriate, although they have been tried. Neither are contract drawings much use; unfolding them does not cause too many problems, but folding them back up certainly does! Candidates do tend to look as though they are trying to wallpaper the room.

There have been many problems with computers, and I note that the rules suggest that

'the use of laptop computers is not recommended'.

This does not mean that you cannot use one, but it does give due warning that you could be introducing undue complexity and unnecessary problems. Some of these, based on several years' experience of the review process, include:

❑ the temptation to use the software to the full, producing an 'all singing, all dancing' presentation. The Reviewers have found that this gets in the way of developing a relationship between themselves and the candidate, and this is the prime reason for the recommendation;

❑ modern display screens are designed to be very directional, so that the person sitting next to you on the train or plane cannot see what you are doing. It is therefore almost impossible to adjust the screen to be seen by the two Reviewers, unless you use a widescreen model;

❏ battery life (no power is available during the interview) is limited, although much better than previously. How many times have you just had a little practice whilst awaiting the interview? Just how much battery life do you have left?

❏ the computer needs to be switched on well before you are called to interview, otherwise you spend an inordinate amount of (your) time waiting. While you do have a stated start time, it is flexible and, luck being what it is, you may be called earlier than expected.

If you do decide to use a laptop, perhaps because you use one for presentations for work and it is a familiar technique, then think it through. Keep the visual content as simple as possible. Do not use animation. Anticipate the Reviewers' first question: 'Did you prepare that yourself?' and give some thought to whether it is the task of a professional engineer to prepare computer presentations. What would you do on the day if the technology failed? Presumably, take a hard copy back-up. If you do, why not just use that instead, and avoid the complexities of a laptop?

Similar remarks apply to those little battery-operated slide projectors; they too, are difficult to use across a table. When used at home, we do tend to sit alongside the viewer to whom we are showing the pictures, rather than on opposite sides of the table.

More appropriate than a flipchart might be something to put on the desk. What that should be is best left to your imagination and creativity. I have seen all kinds of props:

❏ site plans, where a development progressed through a series of acetate overlays, tracing the thought process that went into the design. If you use this technique, frame the original plan in cardboard or balsa wood. It is almost impossible to fit an overlay exactly over the plan when your hands are shaking, unless there is a frame to help;

❏ two- and three-dimensional models;

❏ pieces of rusty metal;

❏ a housebrick;

❏ a disc cut from a suspension cable;

❏ a piece of experimental apparatus;

❏ even a jar of some noxious euent, the lid of which the candidate unscrewed at a key moment, to the amused discomfort of the Reviewers.

There is no requirement to take any visual aids at all. I have seen a candidate use nothing more than a felt-tipped pen to draw on the scrap paper provided. That takes a certain amount of self-confidence, but for him it was a normal method of communication with architectural clients.

The important thing is to make sure by adequate practice that you can effectively handle any props, even when very nervous. Fumbling about in an attempt to get something to work will not inspire confidence.

This brings me to another point. You will, I hope, take documents with you in case specific questions are asked about matters which did not have supporting documents in the submission. If you need to refer to them, do make sure they are catalogued in some suitable order in your briefcase and that you have practised finding them. You do not want to spend great lengths of time bent double behind the desk, with the Reviewers wondering what you are doing! Again, take 'just enough'; be ruthless and discard anything which is not going to significantly reinforce your case – not 'just in case' but vital.

The need for practice

Having decided what to include and how to present it and drafted a script, you must then practise. Even very famous and apparently effortless public speakers invariably practise. I know one such speaker who *never* allows himself to be persuaded to make a speech without having prepared something first, perhaps not specifically for that occasion, but after many years' experience, he has sufficient material and can adapt it to a rough outline in his mind. He also regularly stands back from a full length mirror and critically observes how he uses gestures and movements to reinforce his message. Yet most people believe he is a brilliant impromptu speaker who finds it easy to talk at any time on any matter. I do not think anyone ever really finds public speaking easy, however long they have been doing it. But done well, it can be very satisfying and very impressive.

The ideal place for you to practise is seated at a table with a mirror propped up in front of you so that your image in the mirror is about as far away as the Reviewers will be on the day. Look yourself in the eye, watch how you move and, particularly, see how you handle your visual aids. Speak out loud, pitch your voice at your reflection and time yourself. You will feel very self-conscious, but this is a good thing because it actually causes the adrenalin to flow, thus replicating the anxiety you will feel on the day. Your timing is therefore likely to be about right.

Timing a talk cold or in your head is never successful; everybody goes rather ponderously in their mind, whereas in the real situation you will either put in extra material and go even slower, or you will gabble and finish early! Which scenario applies to you? Only practice in real situations will enable you to find out.

If you intend to use some form of prompt, keep checking that the words on it are what you need; each time you are disconcerted during practice, see whether better keywords or a different layout would help, until your prompt is honed to perfection. Personally, I believe that if you practise hard, you will have no need of a comprehensive prompt; perhaps just a list of the key points laid on the desk on a small card. This leaves your hands free to manipulate any visual aids.

Having done all this, get colleagues, relations – anyone – to listen to you. An ideal opportunity is the Graduates' and Students' Papers Competition, even though the scenario is different – you will stand and use bigger visual aids for a larger audience. Seek opinions, criticism and advice – you do not need to take it all, but it will all help. Furthermore, you will become more confident and begin to relax a little.

Nevertheless, on the day, you *will* be nervous – this is a good thing! Any actor will tell you that you have to be nervous to give a good performance; the day they become complacent is the day they leave the cast. What you must be able to do is control and conceal your nerves and the only way to accomplish this is to practise. Actors and good public speakers learn techniques for controlling their nerves – deep breathing, shaking and relaxing their limbs; there are all sorts of techniques which help. Throughout your previous training, you should have been taking every opportunity to speak in public so that you have developed these skills. Few of you will have done so; as a result, you need to practise and learn now.

Chapter 18

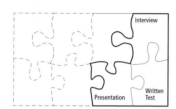

The review day

Everyone tends to worry about the interview and the written work (covered separately in Chapter 19), because these are the first things which are not totally under your control. Five-eighths of the jigsaw is entirely under your control and a further eighth, the presentation, is largely under your control. On that basis, it is difficult to understand why anyone can be unsuccessful if three-quarters of the whole review can be completed without any participation from your Reviewers!

What is likely to happen during the interview? The straightforward and obvious answer is – 'Who knows?'. The Reviewers will by then have a broad plan of what they wish to cover (areas to be explored further) and will have decided on the titles for the afternoon's written work, if that is required.

What they will not have is a list of specific questions which they intend to work through; they will play the interview by ear to a large extent, working from your presentation, their notes of any areas of your experience they wish to explore in greater depth and your reactions to their questions.

Nevertheless, there are many other uncertainties which can, and ought to, be removed. Any uncertainty breeds doubt and doubt fuels nerves, so the fewer uncertainties there are on the day, the better you are likely to perform.

Preliminaries

The first uncertainty is the journey – how and when you are going to get to the venue you chose on your submission form? If you are travelling the night before, have you arranged somewhere to stay? If at all possible, my advice would be to stay with a friend, either in their home or together in bed-and-breakfast nearby (the full rate in most of the provincial venues used by the Institution and those hotels local to the headquarters in London, is usually exorbitant). Being on your own in a strange place late in the evening with no one to talk to can be daunting and confidence sapping; it is not the best preparation for a full, stressful and taxing day.

Whatever arrangement you decide upon, it is sensible to do a reconnaissance. If you intend to travel on the morning of the review, then do the same journey at roughly the same time on a weekday to gauge the traffic or the time on public transport. When stationary in a jam, it is comforting to know just how much longer it will take to arrive when you finally start to move again!

Visit the venue, get the feel of the place. The Institution in Great George Street is in fact quite a friendly building, but you will probably find it very daunting if you visit for the first time on the review day. Have a drink in the bar, either there or in the provincial hotel venue. Find out what you can do (gym, swimming pool, sit in the park, etc.) between the interview and the written work. This period could be as long as three and a half hours, during which you must be able to relax as far as possible; sitting in your car or in a café 'revising' merely makes you more tired and seems inevitably to remind you of all the things you do not know – definitely not the best way to prepare for the afternoon.

Appearance

The clue to the question of what to wear is the same as it has been for every aspect of the preparation – what are you trying to demonstrate – that you are a professional engineer. So look like one!

My only criterion is that you should feel comfortable and enjoy whatever you are wearing; you do not want another uncertainty on the day. Do not wear something which you continually have to tug or adjust because you feel uncomfortable. Do not buy a dark formal suit if your style is somewhat more adventurous, but at the same time remember that most Reviewers are probably nearer the age of your parents rather than you – what would they expect?

Perhaps more importantly, what do your clients or the public expect? Most, I think, would expect a jacket but not necessarily a suit, normally with a subdued tie for men and modest neckline for women; few engineers (particularly the older ones) like or empathise with power dressing or strong fashion statements. Again, it is a question of judgement and the impression you are creating. Avoid distractions of all kinds – rattling jewellery, swaying earrings, a watch that beeps every quarter, anything which is going to become annoying in a tense atmosphere. We have even had a mobile phone ring – the caller wanting to know how the candidate had got on!

I have also been asked (by a man) what the Institution's attitude is to long hair. My answer was, 'We think it's lovely!' from which he took me to mean that he should get it cut. No! I asked him whether he ever represented his organisation to the public and whether they took him and his views seriously. The answer to both was a resounding 'Yes', so he did not need to cut his hair, which in any case was extremely well groomed. Perhaps he did put himself initially at a slight disadvantage with more staid Reviewers – but if that was the case, he overcame any covert discrimination and was impressively successful.

The weather on the day may mean that you have to wear a coat or carry an umbrella. Leave them in the cloakroom to avoid the clutter which would otherwise result as you enter the interview room.

Arrival

There will be quite a lot of activity in the building on the day. There could be as many as two dozen reviews all going on at the same time, so there will be upwards of 60 people directly involved in your 'round' and more from the one which is just ending.

Find the reception desk and book in as soon as you can, before finding the cloakroom (you should know where it is beforehand) and getting rid of everything you do not need. You will probably already have a briefcase and possibly some form of presentation material, so do not clutter yourself further with a raincoat and umbrella.

Remember that the first thing that one of your Reviewers will do, after coming out to the reception area to invite you to join them, is to shake your hand – keep your right hand free! Otherwise the very first visual contact you have with your Reviewer will put you on the defensive, as you fiddle about getting flustered, trying to pick everything up and then changing everything around to shake hands! This is why I recommend you to relieve yourself of any surplus baggage.

This reminds me of another vital point; nerves make us all want to go to the toilet, so make sure you go before, and do not feel the need during, the review! You do not need any unnecessary distractions! Get back and sit in the waiting area about ten minutes before your due time; it is not unusual for the Reviewers to be running a few minutes early.

There is a strained atmosphere in the waiting area at this time, since candidates deal with their nerves in differing ways; some babble continuously, some sit in a corner silently, others feel the need to walk about. Whatever you do, don't try to revise or go through your visual aids once again; it is too late!

In the waiting area, there are usually a couple of senior engineers strolling around, as well as the administrative team: a stand-by Reviewer, there in case of problems and a member of the regional support team. They will talk to you, not with any ulterior motive, but merely to try to relax you. Everybody, from administrative staff to the Reviewers, is willing you to do well, and will do their best to assist you to do so.

One of your Reviewers will come out to the waiting area to get you. From that first moment of contact you are being observed! At least look as though you mean business, even if you feel awful. It may interest you to know that the Reviewers have usually only just met each other, so they are nearly as unfamiliar with each other as they are with you. Hopefully you have remembered both of their names (which you were given for posting your documents) so even in the heat of the moment, you should not forget who they are. If one forgets to introduce you to their colleague, then introduce yourself – 'You must be . . . ?'. At some point during the interview, try to refer to each of them by name – it suggests you are in control of yourself and the situation.

During the short walk to the meeting desk, your Reviewer will talk informally, probably telling you a little of their own background and asking whether you have had a reasonable journey – anything to try to get you to relax. Talk to them! Don't just say, 'Yes, thank you' but have something in mind – 'Yes, but did you get stuck in that sewer replacement job in the High Street?'. Such an answer shows that even when under pressure you are interested in what is going on around you and had the foresight to anticipate possible delays – inspiring confidence. Even casual conversation is creating an impression and commencing to develop a relationship which *must* blossom and reach fruition in the next hour!

Interview scenario

This is your opportunity!

It is not for the Reviewers to find out your competence, but for you to demonstrate it.

The scenario is that you are approaching two existing members of the Institution and telling them that you have become their equivalent – a professional engineer. Their response is 'Go on then, prove it!'.

Your table will be one of several, arranged round a large room in such a way that, as far as possible, candidates are not in eye contact. It sometimes happens that one of your senior staff, or someone you know, is acting as a Reviewer on the same day; if you know that this is a possibility, do inform the Institution and strenuous efforts will be made to ensure you are not in eye or sound contact with them. That is the last thing you need on such a stressful occasion!

Your two Reviewers will sit opposite you at an average sized table (around 1.5 to 1.8 m × 1.0 m); there is not a lot of room after you add two or three drinks, two sets of open submission documents and some scrap paper and a pencil! As far as is practicable, the Reviewers sit facing into the room, with you facing the Reviewers with a wall behind them.

If no drink has been provided for you, ask if you may go and get one – there is usually a table somewhere in the middle with water and other soft drinks supplied. Nerves will make your mouth dry, so be prepared – always look in control.

Observers

There may be a fourth person sitting at one end of the desk, within your eye range. They could be an observer from the Institution or the Engineering Council, as part of the quality assurance processes, or a new Reviewer learning by shadowing. They will take absolutely no part in the proceedings and have no influence on the outcome; you do not need to include them in your eye contact if you find it awkward.

Usually you will be asked beforehand by the administrators whether you mind the observer. There is no reason why you should not say 'Yes' if you feel particularly nervous; it certainly will not affect the conduct or result of the review. If you object, the administrators will try to move that person somewhere else.

Presentation

Picture the scenario when deciding on how you will make your presentation and particularly when you practise, as you must if you are going to reduce all the uncertainties which otherwise surround this aspect of the day.

A projector and screen or video are not deemed appropriate for this intimate setting; you do not need a pointer or light pen and anything you use to illustrate your work should be to the scale of the setting – A3 maximum, certainly not A0 and in my view preferably A4. After all, there is an audience of two, not two hundred! The scene is informal and quite intimate. You will have to raise your voice slightly to overcome the ambient noise levels, but do not throw it as though you were talking to thousands; it is inappropriate and a distraction for everyone else.

How are you going to start? Do not introduce yourself again; they know who you are already. Too many candidates start in a loud voice 'For my presentation today, I am going to...'. Not appropriate! Might I suggest, 'I know you will have read my Project Report, so for the next 15 minutes, I am going to concentrate on (whatever it is you have decided to expand on)' or something similar. Remember the old adage: tell them what you are going to tell them, tell them, and then quickly summarise what you have told them.

Make sure, if necessary by asking, that both Reviewers can see your visual aids. So often a candidate puts a flipchart on the left corner of the desk, forcing the right-hand Reviewer to lean across to be able to see properly. Move your chair so that your visual aids are centre stage during this formal part, but remember to move it back as you finish and you become the centre of interest.

I personally would not stand throughout, although I have seen candidates become so enthralled in aspects of their work that they have momentarily stood up to demonstrate a point. And this really is the key to success. Even though you will have practised many times, you must not give the impression of 'going through the motions'; your Reviewers will be genuinely interested, so do not dull their interest by boring them. You must inspire them with your enthusiasm.

At the end of your presentation, give your Reviewers a clue that you have come to the end; they may otherwise think that you are merely pausing to gather breath or have momentarily lost the thread. Close your presentation material, move your chair back to centre stage, say, 'That is the end of my formal presentation. If you have any questions, I shall be pleased to answer them'.

Make sure that you stay just within the time limit; anything longer will start to irritate the Reviewers, who are aware of how much ground remains to be covered in such a limited time. You are unlikely to get away with the tactic of extending your talk to reduce the questioning!

Interview questions

The questioning which follows the presentation is really a series of prompts to set you off talking again in a direction determined by the Reviewers, so do not give monosyllabic answers. They wish to see an engineer's mind at work, talking through problems and difficulties and discussing the judgements which have to be made every day of our working lives. The Reviewers manage the interview so that you talk for about 80 per cent of the time, and they then use the remaining 20 per cent to direct you towards the areas they wish to explore further.

Do not be disconcerted if you are asked questions you cannot answer. I have discussed this with Reviewers, who have told me that occasionally, they do not know the answer either! What they want to see is what comes into your mind, how you tackle the question, what you would need to find out to arrive at an answer – in other words, what you do every time you are faced with an apparently insoluble problem at work.

As one Reviewer described it, 'Engineers stumble towards a solution in a relatively organised way! I would like to see the candidate doing just that'. This is why I counsel that the review is very different from the conventional examinations you have experienced up to now, where you have been expected to give authoritative factual answers to any questions you have been asked. Engineering problems cannot be resolved by such black and white decisions. The Reviewers are anxious to see that you are capable of tackling complex problems, understanding the implications and using your judgement.

If you feel yourself coming under pressure, it is because you are not delivering what the Reviewers need in order to pass you. They are not trying to catch you out, but are attempting to get you to demonstrate some attribute which, as yet, they are unsure of. Try to detach yourself; become a third party and mentally step aside to see what it is they really want. After all, you have convinced yourself and your sponsors that you are capable and competent, so you ought to feel confident that you are able to persuade these two fellow (or peer) engineers.

One of the things which really surprised me when I first became involved in the administration of the reviews, was how much laughter emanates from the room, particularly towards the end of each batch of interviews. Looking back to my own interview, for which I was extraordinarily well prepared by my Chief Engineer, I recollect that we had quite a jovial discussion on one particular aspect of my experience not mentioned until the very end – it turned out to be the essay topic!

The Reviewers really do enjoy meeting good candidates; for them a good morning consists of *successful* reviews. This is why most of them volunteer. They do not enjoy interviews where progress has been 'like drawing teeth' or meeting someone who was inadequately prepared or not up to standard. You would probably not enjoy that sort of interview either, so do not tempt fate.

One last point about the morning – your two Reviewers will be delighted to relieve themselves of their copies of your submission, so you will be coming out of the room with three piles of documents. It is rather nice to be able, at this stage, to draw out a plastic bag to put the documents into; it shows you had anticipated that they would hand them back and everything is under control.

Beware the quick throwaway question just as you relax – you are still under observation until you leave the room.

The interval

Try not to draw any conclusions from the way you felt the interview went. First, you cannot really remember – it will all just be a blur! Second, the Reviewers are unlikely to have revealed their thoughts and opinions, either about you, your ideas or your experience; so any conclusions you come to are based on very little factual evidence. I have known candidates to burst into tears or become extremely angry, convinced they have failed, when in fact, they did really well. Others have come out feeling very confident, only to get a letter in due course pointing out where they went horribly wrong.

You may, like a boxer, feel battered and bruised; just try to return for the next round – the written part – convinced you can deliver the knockout blow! This is where you need your friend again; like a good second, they need to bolster you up for this final round.

There is nothing to be gained by shutting yourself away in your car and mulling over everything that went wrong. Because this is what

I fear you will do. It is the worst possible environment and the worst possible isolation. Even though I have been tutoring now for nearly 20 years, by the time I arrive home late after a long, lonely drive, mulling over what I said, which I should not have said, what I could have explained better, and what I forgot altogether, I have convinced myself that 'Today it was a disaster!'. I do not know why the human mind works in this way, but it does seem to for most of us.

Far better to join other candidates and talk, not about the interviews but about your work and experiences, last night's match, the latest music or fashion craze – anything to take your mind away from the morning. It all helps you to relax in preparation for the last burst of adrenalin; like any good athlete, you must pace yourself.

Chapter 19

The Written Test

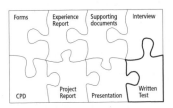

Purpose

This final part of the review brings the whole jigsaw to completion, locking all the pieces together. It is often said that this part is the most common cause of failure, but, analysing the figures in rather more depth, it quickly becomes clear that if the written work is removed from the statistics, the review pass rate hardly alters. This means that those whose written work is inadequate are generally also found wanting in other aspects, or pieces of the jigsaw. The implication is therefore that all the main pieces of the jigsaw will become connected to this final piece – all your experience and expertise from the report and supporting documents, and your CPD. I have even known instances where the Reviewers felt that a candidate did not give as good an answer as they were capable of during the interview and so asked the same question again for the Written Test.

In my search for guidance to help potential candidates, I eventually found what I consider to be the most useful and concise guidance:

> 'carry on your discussions with clarity and power and rigorousness, in recognisable sequences of enquiry, discovery, expansion, challenges and conclusion, all conducted with reason and addressed to any subject that takes your fancy.'
>
> (Francis Bacon)

Francis Bacon was one of the first Britons to relate scientific discovery to practical problems through meticulous research and progressive experiment. It is perhaps fitting that, as an early exponent of what we would now describe as engineering thought, he should also have produced a description which I believe is of inestimable use to civil engineers.

In the context of the professional reviews, I believe that it is reasonable to divide the types of question you could be asked into four broad categories:

(*a*) *Factual* – e.g. 'Outline the stages in the development of a design for a (bridge/dam/leisure centre ...) from the initial brief to the issue of working drawings, giving examples from your own experience.'

(*b*) *Expository* – e.g. 'List the records which should be kept by the Resident Engineer responsible for the construction of a waste-water treatment works. Describe the content and function of each item.'

(*c*) *Argumentative* – e.g. 'Discuss the advantages and disadvantages of using concrete additives in the construction of motorway pavements.'

(*d*) *Visionary* – e.g. 'Engineers are born not made. Discuss.'

In general, although not invariably, the first two types of question are usual for the Associate or Member Professional Reviews. You are expected to be able to communicate, in writing, facts and procedures with which you should be familiar, since the questions are based on your direct experience. This type of essay should enable you to demonstrate:

❏ the ability to marshal thoughts quickly into a reasoned order to a tight deadline;

❏ that you are decisive, clear thinking, and can put a reasoned and logical description together quickly;

❏ that you are able to explain something with which you are familiar to someone who is not;

❏ that you have an incisive ability to 'see the wood from the trees', without becoming immersed in repetitive or irrelevant detail.

These are the characteristics of any competent professional engineer.

The latter two styles of question ask you to state and justify your own views and opinions and are perhaps more representative of the questions set for the Chartered Professional Review. There is no 'correct' answer to such questions. Indeed the Reviewers are expressly told that they cannot mark a candidate down merely because the views expressed differ from their own, with one very important proviso – that the views are not mere prejudice, but are ones which could reasonably be developed from the experience of the candidate. These latter types enable you to demonstrate all the above listed qualities, with the addition of:

❏ independence of thought and opinion,

❏ a professional attitude, integrity and honesty,

❏ judgement for the benefit of the general public good,

❏ wide knowledge and understanding of current aairs,

– all attributes of a leader of the profession – a Chartered Engineer.

These scripts are not in any way like the written examinations taken during an academic course; they are *not* a test of knowledge, but a test of your ability to communicate as a professional engineer. Knowledge is used to *support* your responsible reasoning and arguments (hence the marking of 'relevance' – of your knowledge to the question and to your opinions).

Method of approach to the Written Test

Consider the keywords which come out of Bacon's definition, coupled with the notes in the Institution guidance, and compile them into the framework or pattern:

Enquiry	What does the set question mean?
	Why has it been asked?
	What does it imply?
	What lies behind it?
	How are you going to attempt or endeavour to answer it?
Discovery	This is the thinking part of the process, where you explore all the relevant facts and knowledge you can

think of. You will be searching for references and background material. Then you must marshal your thoughts into a logical sequence.

Expansion   Where does the question lead? Can you determine the full extent of your answer and develop a line of reasoning or argument?
Can you extrapolate from the immediate answer to the question to broaden out your discussion sufficiently?
Do you have the relevant knowledge to support your statements or opinions?

Challenge   Can you provoke thought or debate in the Reviewers' minds?
State your own views positively and concisely.

Conclusion   Summarise your own personal thoughts or arguments.
Do not introduce new material. The conclusion must be a logical summary of what you have already said.
If you are running out of time, then list the outstanding matters you hoped to discuss.

In effect, your written work leads the reader from the question, through a reasoned and logical discussion or explanation which they should follow and accept, to a conclusion which sets out *your* views on, or understanding of, the answer to the question, with which (hopefully) the reader will then agree, or which will provoke them into thinking more about it themselves.

Newspaper editorials are a useful source of ideas on approach; note how, while the content differs from day to day, the format or 'technique' is nearly always the same every day. One useful technique is to compare a news item as described in a broadsheet to the same item in a tabloid. The journalists who work for the tabloids are experts at minimising the word count but still getting the story across. This is what you need to do.

You will be familiar with the phrase, 'Tell them what you are going to tell them, tell them, then tell them what you have told them.' I do not think this is particularly relevant in this situation, where time will not allow much repetition, but, nevertheless, I believe you should tell the Reviewers how you are going to answer the question (which may identify matters you will not have time to cover), answer the question and then draw the whole thing to a conclusion by summarising what you have just said.

Preparation

The purpose of the Written Test is for you to demonstrate your ability to communicate in writing through scrutiny of:

❏ the way in which the content relates to the question – this includes the factual content and the way in which various aspects of the subject are analysed, compared and contrasted and conclusions reached;

❏ the structure – there must be a clearly identifiable *introduction* displaying the main aspects of the topic; a *development*, in which the topic is considered in detail and the main and subsidiary arguments set out; a *conclusion*, in which the threads of the discussion are drawn together and conclusions reached;

❏ the quality of the English – this relates not only to grammar and spelling, but also to sentence and paragraph construction and style.

Technique

The Reviewers look upon candidates as potential senior managers who will be required to present reports or advice to laymen, so a reasonably high standard is expected. Past inability to express ourselves clearly (except, of course, to other engineers) has contributed to a lack of public confidence. It is worthwhile therefore devoting some time to perfecting a technique; there are several complementary ways in which this can be achieved.

❏ Attendance on a course – there are many throughout the country, but look at the format carefully; you must consider carefully what you need.

❏ Is a concentrated four-day course better than 26 weekly evenings?

❏ Is your weakness the use of language, grammar and syntax or breadth of knowledge?

❏ Would it help if your practice essays were marked?

Thomas Telford Training has a very good distance learning package which enables you to work around your normal work commitments – particularly useful if you are site based.

❏ A selective programme of reading – this includes publications relating not just to technical matters in which you are involved but to your profession, management generally and the political, financial and environmental framework within which you work.

❏ Development of a critical awareness in the workplace, not merely of your own involvement but of the whole environment within which you operate.

❏ Suitable practice under 'examination conditions'; this is by far the most important. As an invigilator, it is clear to me that too many candidates are *physically* unfit for the Written Test, i.e. they cannot write for two hours without discomfort. After all, when was the last time you wrote *continuously* for that length of time – during your finals? If you are thinking about cramp and discomfort, you cannot be concentrating on what you are writing!

The Written Test questions (two, from which you must select one) are set from your submission; it is expected that you will know at least enough to answer either of them. (If not, the Reviewers have found you out – your report has misled them!) Do read both questions and make sure you understand what the Reviewers want; it helps if you underline the keywords. Do you understand what is required by 'Discuss', 'Describe', 'Comment' etc.? But remember, unlike much of what you have written over the last several years, this is not primarily a test of knowledge; it is a test of how well you can communicate your knowledge to a layman (that is, someone of equivalent intelligence but without your specific knowledge). In this respect it differs markedly from most, if not all, of the written communication you carried out during your academic education.

Planning

Do not choose one of the questions without a few minutes' thought. It is surprising just how many scripts start on one question and, after a few lines, the candidate has crossed it all out and started again on the other. Not only have they put themselves at a considerable time disadvantage, they have also given their confidence a nasty jolt.

Do not start the actual writing without adequate planning. Think about the answers to both questions for about five minutes, then move on to develop a plan for the better one. For 1000 words, 20 minutes should give enough time to construct a plan. This may seem excessive, but, with practice, most people can physically write 1000 words in less than an hour, thus giving plenty of time to check the product at the end of two hours.

At the start of the assignment, 25 or 30 minutes will seem like a lifetime – all your instincts will be screaming at you to start and everyone around you will be scribbling frantically. But control yourself, secure in the knowledge that previous experience under similar conditions has made you confident you can write sufficient in the remaining time.

Planning is itself assessed – any drafts, notes, plans, etc. should be inside the cover of the booklet provided, if you are to receive any credit for your planning; this is why no scrap paper is allowed. Once you have decided on a plan, stick to it and finish it – your Reviewers will not be impressed if you cannot even fulfil your own programmed content. Monitor progress by deleting each idea once it has been incorporated.

Ideas will spring to mind in a very haphazard way when you begin to think. It is important to jot down every idea in note form as it occurs, however random and irrelevant it seems. The 'braindumping' after the initial blank panic at the start will, generally speaking, begin with the answers to the question, followed by 'less important' material to provide enough to write 1000 words. The temptation is to follow the same logical process, answer the question in the first few pages and then proceed to justify the answer; the result being that the essay tails away.

It is better to start by filling in the background first, arriving at the answer at the end – i.e. invert the list. In this way you lead the reader to a firm conclusion. But beware! Do not pad out the start with irrelevancies; how many times have I seen a Reviewer's comment somewhere about page three, 'answer starts here' – not a good omen!

Once all your ideas are recorded, you can assemble them in a logical order; there are several ways of doing this:

(*a*) simply numbering each idea in a sequence and then rewriting in the correct order – the danger here is that it is relatively easy to miss one out;

(*b*) lettering (or using a colour code) to identify each idea to its appropriate paragraph and then numbering each within the paragraph, finally rewriting in the correct sequence;

(*c*) spider (spoke) diagrams where ideas radiate outwards from the central question, each being assigned to the most appropriate 'leg' – each leg subsequently becomes a paragraph;

(*d*) 'noughts and crosses framework', achieved by dividing the answer into three parts, each containing three paragraphs. Each idea as it occurs is put into the most appropriate 'box'. This is a highly structured technique which requires practice, but is very useful for writing short reports.

I also know of another method known as 'mind mapping', used by the Open University and on which many books have been written. It is a more sophisticated 'spider' system. Use whichever system you feel comfortable with.

Do not get carried away with one particular aspect to the detriment of others; try to achieve a *balance* in both content and depth of treatment. Planning methods (*d*) and (*c*) and to a lesser extent (*b*) clearly highlight any such imbalance since one square or leg will become overloaded. If it does, reconsider the distribution, consider adding another paragraph or perhaps you are entering into too much detail on this aspect.

Planning gives you the structure of the body of the assignment, especially the paragraphs. A paragraph is a collection of sentences all on a particular theme. Change the theme – start a new paragraph. The use of paragraphs demonstrates control over ideas and their expression. A good answer contains paragraphs of variable length, each containing material on one aspect only. The paragraphs themselves should follow a logical sequence, avoiding abrupt changes in direction which disconcert the reader.

The booklet provided by the Institution has a rather unexpected form; the back cover is A3 folded in. The covers themselves are different colours for different purposes and the invigilator will ask you to check that you have the correct one for your particular needs. But the purpose of this A3 cover is for you to do the brainstorming on the left-hand side and prepare the plan on the right-hand side. Left open, your plan is then visible all the time you are writing, to the right of the main booklet. Not so convenient for left-handed people, but still useful.

Format

The Reviewer will also consider how effectively you introduce and conclude the answer. The opening sentence must attract interest and make the reader want to read on; a short, sharp first sentence is the most effective. It can be a good idea to rewrite the question and outline your approach to it – this helps to ensure that you do actually answer the set question and not the one you thought (or hoped) it was.

There must be a conclusion, a 'summing-up'; even if you are short of time, at least a 'one-liner'. Avoid dullness and assumptions; do not introduce further information; do make sure that your development does actually lead to the conclusion you have reached! The conclusion must be positive – do not let the essay just peter out.

The Reviewers are looking for understanding supported by facts. Do not be afraid to express original ideas and opinions (as long as they are sensible and you justify them) rather than regurgitating stereotyped 'popular' views. Present points in an orderly, uncomplicated way to demonstrate logical, clear and, if possible, original thought. Substantiate each point with facts or figures, wherever possible taken from your own experience. You have met your Reviewers by then – do not fall into the trap of subconsciously writing for them, knowing as you do that they are already familiar with the subject – the result will be superficial and unacceptable.

Clarity and presentation

Make the script look organised and authoritative. Practise handwriting if yours is rather scruffy and difficult to read. In any case, when was the last time you wrote for an hour without a break – you need the practice! One useful method is the Quarterly Report; of the same length but different format, once you have decided on a plan you should be able to write each one in an hour (though I have found few who have!). Try to keep the handwriting consistent throughout, so that once the Reviewers are used to it they can read it easily. If all of this is difficult for you, then consider the use of a computer, but do read and digest my advice at the end of this chapter.

Indent paragraphs in a consistent way – begin on the next line (no space between paragraphs) about 3 cm from the left-hand margin.

Remember that the Reviewer is reading your work in their own time and may well be tired; make it clear, readily understandable and enjoyable, but avoid making jokes – they will inevitably be taken the wrong way!

Avoid lapsing into jargon or slang; do not use words like rebar, shutters, dumper or lab. This is not easy, because we use both jargon and slang every day in discussions with other engineers, but these are not considered appropriate in the written word.

Under no circumstances use several words where one would do because you think the script may be a bit short; this will only draw the Reviewers' attention to the deficiency and probably annoy them! If you use abbreviations or acronyms, on the first occasion write them out in full, followed by the acronym in brackets.

Practice

To be successful in the Written Test, it is vital to practise under the supervision of someone (not necessarily another engineer) who can comment not so much on the content but on the 'readability' of your work. They need to be asked such questions as: 'Do you see what I am getting at?' 'Can you follow my argument?' 'Does it make sense?' 'Did you enjoy reading it?'. Such a person need not be an engineer; in fact sometimes it is a positive benefit that they are not familiar with the subject that you are writing about. They will ask questions which cause you to consider whether or not you did actually write what you intended, or whether there is a better way of explaining that particular matter.

Having watched a large number of candidates writing and discussed the problems with them and the Reviewers, it is absolutely certain that there is no substitute for practice – not just in writing answers, but in writing them under severe time constraints. Anyone can write an acceptable answer in a fortnight; it takes skill, time management and clear, quick thinking to do it adequately in a couple of hours. I once received a long letter from a friend which concluded, 'I am sorry this letter turned out to be so long. I did not have the time to write a shorter one!'

I become concerned when I visit essay groups at the amount of time and effort spent collecting knowledge – generally far too much to regurgitate in the given time. All that many of these groups are in fact achieving is to burden their members with an additional problem – too great a choice of available information! By all means read and discuss around the subjects, but then distil this information into key points, which can be introduced into many of the questions. It is my considered view that most essay groups would be much more effective if they concentrated more on practice than on collecting information.

The other concern I have is the predilection for 'model' answers. Now that the Written Test is 'open book', and the questions must be specific to your particular experience, it is highly unlikely that any Reviewer will ask you a stock question. If you have never been involved in joint ventures, then you will not be asked a question about them. In one extreme example, a candidate with a background entirely in water engineering answered a question using examples from highway work. Obviously he had swatted up model answers and was disconcerted when he failed the written test on knowledge and relevance! Even within a group from the same organisation, the answers will be different, reflecting the differing experience of the members. Model answers may give you a lead and a general shape to an answer, but you must put your own perspective on them.

You need a working knowledge of the whole range of civil engineering, not merely a high degree of specialist technical knowledge, so that you can relate your expertise to the environment in which you work. You also need to immerse yourself in well written, well structured papers, magazines and books so that you subconsciously absorb good style and techniques.

English as a foreign language

The language of the Engineering Council is English, so anyone wishing to register with them as CEng, IEng or EngTech must demonstrate that they have a reasonable facility with it. The Institution can and does offer membership overseas, where the Written Test is done in the candidate's own language, but then cannot recommend the candidate to the Engineering Council. Since the Engineering Council's jurisdiction only extends to the UK, this is usually of no relevance.

You are trying to prove you are a professional engineer and, as such, any reports or documents which go out from your organisation would be expected to be of a consistently high standard. So your reports and submission must be perfect! As a responsible engineer, you are expected to have taken any necessary steps to ensure this, whether English is your first language or not.

The standard expected for the written work in the afternoon of the review is different. The documents state 'a final draft'. In other words, not perfect, but tidy and logical, capable of being edited by someone without your technical background. A few crossings out might be acceptable, but great circles around paragraphs to move them elsewhere are not – they demonstrate a lack of foresight and a muddled mind.

Again, your curiosity should have been aroused by the wording in the official booklets, where the requirement states 'in acceptable English'. 'Acceptable' is perhaps an unexpected adjective, but it allows the Reviewers the freedom to make a judgement – acceptable in the context of the background of the candidate. Someone who has come to this country relatively recently may not be as competent in the use of English as someone who was born here. The Reviewers are seeking to confirm that each candidate is capable of communicating in writing in their particular context.

This exercise of judgement also applies to dyslexia, provided that you are able to inform the Institution at the time of submission by supplying a certificate defining your problem. There is not usually any dispensation on the time allowed, but all written work by dyslexic candidates is automatically referred to a special panel with access to appropriate medical advice.

Computer use

You may be surprised that I have left this to the end of the chapter, but everything I have said so far about the Written Test is applicable whether you use a portable computer or not – which ought to give you a good idea of how relevant I think a computer is for the Written Test!

I have discussed why candidates wish to use a laptop with many of them. Some say it is because their handwriting is not legible when under pressure, and I think this is a valid reason.

I am not at all convinced by another common reason: that candidates believe they can write faster on a laptop than they can by hand. I think that in many cases, this is merely an attempt to mask an inability to marshal thoughts and mentally compose a logical discussion, before transmitting them. How are such engineers going to manage when offering verbal opinions in a public arena, or worse, in a court of law?

If you believe that you are going to be able to mask any deficiencies by resorting to the frequent use of cut-and-paste, then I fear you may well run out of time or produce a disjointed script. The Reviewers say that they can usually quickly identify such a technique because the script just does not flow.

If you seriously believe that you are going to have the time to search your hard drive to collect relevant facts and information about the question, and then compose them into some sort of coherent answer, then I think you are deluding yourself on several counts:

❏ your Reviewers will ask you questions to which you ought to know the answer anyway, so you should not need the reassurance of comprehensive details stored in the memory;

❏ you may well fall into the trap, mentioned earlier, of actually having too much material to transmit in the time available, giving you the added diculty of making choices under pressure;

❏ the Reviewers know that you intend to use a computer, and so will not ask any questions to which you may have stored a previously compiled model answer.

Are you, in reality, attempting the review too early? Your apparent dependence on stored material does perhaps suggest that you personally doubt whether you really do understand what you are doing or the wider implications of your work!

So do think hard about why you wish to use a personal computer and ensure that your reasons are valid.

Chapter 20

The aftermath

You have completed the jigsaw to the very best of your ability by about the second week in October or May (in the UK) at the latest. You now face a prolonged wait for the result, which will be published either towards the end of November or in June. It is perhaps difficult for candidates to understand why it takes so long.

When you leave the review centre, the system really swings into action. The literary efforts, amounting to as many as 40 scripts each day for a month, are sorted, parcelled up and posted (generally that same night) to the least experienced Reviewer (the one who probably came out to make the initial introduction at the start of the interview). They will read and mark your work before posting it off to the Lead Reviewer, who in turn will read the script and then confirm the marking. At this stage they will probably confer by phone either to compromise where there are discrepancies in the marking, or to discuss the final completion of their paperwork about you, most of which they try to complete in the short time between each interview. Obviously they now have to add the result of the written work and confirm the tentative conclusion which they arrived at after your interview. Remember that they are all doing this in their own time and fitting it into an already busy schedule of work so there are bound to be some delays.

Your written work, together with the Reviewers' forms, their verdict and a draft of the letter to be sent to you if you have not been successful, are then all parcelled up and sent to the Reviews Office. As between 600 and 800 results arrive, the office staff prepare all the letters for signing by a senior officer, and despatch the random samples which will be put before expert panels for review as part of the quality assurance processes. At this stage, there may also be candidates with dyslexia whose written work is automatically reviewed by a panel. These panels are specially convened and again are dependent on a number of volunteers fitting mutually convenient dates into their diaries.

Meanwhile, the staff file all the paperwork and examine any responses to the announcement published in *New Civil Engineer* some time before, where the membership is asked whether anyone has any reason why the listed names (including yours) should not be admitted to the Institution, provided they are successful at the review. Every draft failure letter is being examined by another panel for any inconsistencies before being printed and signed.

I hope you can now see that everything possible is done to keep the suspense time to a minimum. This is why you will definitely not get any reply to a query about the results. All such a phone call or enquiry will do is delay the process for everybody – so do not be tempted!

The unsuccessful review

There are all sorts of myths about the result envelope, with people convinced that you can tell the result from the size of the envelope. Do not be misled, just open it. If you see the dreaded words 'I regret to inform you ...' that is probably all you will read for some time – all that effort and hard work was in vain. This is a devastating blow. There is little anyone can do to reduce its impact; the Reviewers are urged to tell you what you did correctly, what experience was good and go on to tell you why they reached their decision, but the initial shock will prevent you from looking at this rationally.

You will then go on to read the reasons and will become increasingly disillusioned. It is highly likely that you will disagree with what you think the letter says and feel that it is unfair. The temptation will be to march down the corridor and berate your Supervising Civil Engineer or line manager about the inadequacies of the Institution or pick up the phone and talk belligerently to your Regional support staff. Worse, you may even be tempted to write a strong letter to the President! The one thing you definitely should never do is phone the Reviewers; this is considered totally unprofessional and completely out of order.

Recovering the situation

I strongly counsel caution. Wait! Wait until your initial surge of disappointment, even anger, has dissipated. You will then feel utterly dejected. It is at this stage that you need wise counsel, something which the ICE Regional staff are in a very good position to provide. They see many letters (you have seen only this one and your colleagues hopefully very few more) and they also know most of the Reviewers and their styles of writing. They also have access to advice and guidance from their colleagues on matters of detail with which they might not be familiar. So they will be able to give you an insight into what has actually been said – 'reading between the lines' as it were.

The phrase which appears most frequently is 'You failed to demonstrate . . . '. Your reaction may well be 'But it's obvious!'. Is it? Or are you expecting the Reviewers to infer something which the regulations state has to be *demonstrated* and which has not actually been spelt out in your reports? You and your sponsors *know* what you were trying to write; the Reviewers can only see what you actually wrote down. Did you actually say what you meant?

You may feel that you were not given adequate opportunities during the interview to display your expertise; you must ask the question 'Why?'. Had the Reviewers come to the conclusion that it was not worth pursuing this matter? Your interview should not have been a cross-examination – poking relentlessly at a weakness may only cause a candidate to become defensive and is unlikely to allow them to present themselves favourably, so most Reviewers change the subject. Could you, in fact, have demonstrated the point in question or have they actually identified a real weakness?

Very often, the letter concludes by suggesting that before you submit again, you should seek advice from senior engineers, including your Regional support staff. This is good news, because it suggests that your Reviewers actually believe you stand a fair chance of success if you just rectify the things which they have identified.

Rectifying any problems

There will be one of two outcomes from this analysis of the detail of the letter – either you did 'fail to demonstrate' or the Reviewers actually did identify a weakness. The former can fairly readily be rectified in the next submission, while the latter will take some time. In extreme cases, it may be that they have exposed fundamental weaknesses, either personal or in your experience, which make another attempt unwise. They may suggest that it may be more suitable for you to divert to a different, more appropriate review, certainly for the time being. This is a formidable personal decision to have to make and I urge anyone in this difficult situation to discuss it with as many informed people as they feel able. Do not keep mulling it over yourself – you will not resolve the problem, it will merely loom larger.

Any small shortfall must be put right by further experience; it may be that the Reviewers have given an indication of how this could be achieved. You need to talk it through with key personnel in your organisation, perhaps engineers who sponsored you, to identify opportunities for suitable experience which might become available soon to rectify this weakness and how you could take advantage of them. Perhaps the problem was with the written work, in which case it may be advisable to enrol on an appropriate course. This could be one of the many specifically aimed at the reviews, or it could be a more fundamental course on the use of the English language.

The revised submission

You need to review your entire submission to see whether the various documents can be adjusted to cover the weaknesses perceived last time. You will have had, in the interim, further experience, which might be more appropriate for the Project Report; in any case, your Experience Report will have to be adjusted so that you can include this extra experience.

Every review is a complete entity.

Do not make the fundamental mistake of believing that you have passed most of the review and now only need to rectify the few shortcomings, either perceived or real. You are taking the whole review again – with one important difference. This time the Reviewers will have an additional document – a copy of your failure letter from the Institution. This puts you at a slight disadvantage, so you must get back to the status quo as quickly as possible.

To do this, I believe you must answer the letter which has been copied to the Reviewers. I think it entirely appropriate that you write to each of them (as part of your revised submission), explaining precisely what you have done, in the way of additional experience, Continuing Professional Development and training or in rewriting the submission and amending the supporting documents, to rectify the faults (perceived or otherwise) found last time. This is a totally professional approach; otherwise they will not know what you have done to rectify problems and you will have to try to explain during your interview, which just shortens the time available for them to find out whether you now comply with the requirements and really are a professional engineer.

Chapter 21

Technical Report Route (TRR)

This route is almost worth a book on its own. In an ideal world, perhaps a separate volume would be the best solution, because it is significantly different from all the other ICE reviews. Keeping it separate would accentuate the differences. This is why I have left this chapter until the end.

Like the other chapters, however, this one again concentrates on the philosophy and purposes; you are expected to pick up the rules and regulations from the Institution's publications. Full details can be downloaded from the ICE website.

Eligibility

The TRR to membership is for those engineers who have progressed to roles normally filled by either Chartered or Incorporated members, but who lack the necessary academic qualifications and are unable to top theirs up.

There is a mistaken belief that it is the route for all those whose age exceeds the minimum requirement; this is not the case and in fact would subject many, who have already satisfied the academic base, to a more onerous route than is necessary through the more conventional reviews.

There is an equally erroneous minority belief that anyone with the requisite years of experience is automatically eligible to pursue this route. Look again at my first paragraph, where it is clear that unless you have made considerable career progression and taken increasing responsibility, so that you are virtually operating in the role of a professional engineer, then you are not eligible. This is not a back door route; it is onerous and rigorous.

So the very first thing to do is to decide which is the most appropriate grade of membership for you. Compare your current role with the criteria referred to in Chapter 3 and make an honest judgement about yourself. One of the reasons why there is a preliminary sub-mission for this route is that far too many people seem to believe that merely because they have a lot of experience in civil engineering and have reached maturity, they have a chance of success. This is just not true!

Far better to persuade yourself at the outset of the most suitable and realistic goal than to aim too high and be told, after a large amount of effort and a considerable period of time and several interviews, that your experience has not developed the required levels of ability. There is no room for potential, as there might be for a younger engineer; at the time you submit you must have, to all intents and purposes, become a professional engineer.

The second thing to do before you start in earnest is to make quite certain that your academic qualifications are not acceptable and that you therefore cannot utilise the conventional, and rather less onerous, routes. To some extent this depends not only on what your qualifications are, but also on exactly when, and perhaps where, you achieved them. The rules are complex and you need to get informed advice and guidance from the Education Depart-ment at the Institution. Many overseas qualifications are (or can be) ratified, so even if your degree or diploma is not apparently eligible, do check. You could save yourself a significant amount of effort.

So, for this review, you have two things to prove – first, that you are operating at a level of responsibility commensurate with that of the professional qualification you seek and second, that you have achieved a standard of technical and academic competence comparable with that possessed by an academically qualified engineer in the same position of responsibility.

It is this second requirement which seems to cause the greatest difficulty in the progression, mainly, I think, because candidates concentrate exclusively on proving their professional responsibility. This is, after all, probably the thing which makes them want to become professionally qualified in the first place.

By and large, your Experience Report is the vehicle to prove that you are operating at a sufficiently high level of responsibility, while your Technical Report will demonstrate your understanding of technical principles. It is vital that you keep this distinction in mind throughout your preparation of the documents.

If you have the competencies set out in the appropriate Appendix A of the following *Routes to Membership* and the following **indicative** years of increasingly responsible experience listed by the Institution, then TRR is an option:

	IEng	CEng
Accredited BEng/BSc or equivalent	–	7 years
Accredited HND/C or equivalent	5 years	10 years
Approved ND/NC	10 years	–
No appropriate qualifications	15 years	15 years

Before starting on this route, engineers are required to submit a c.v., a synopsis of their proposed Technical Report and the name of a mentor, so that ICE may ensure a reasonable chance of success, thus avoiding abortive work and unrealistic expectations.

Do not produce the synopsis before you at least have a draft report. I hate to think of the number of candidates who have written and submitted a synopsis and then found they could not write the Technical Report to back it up! This synopsis is critical and is frequently a stumbling block for a large proportion of candidates. So again, it needs careful construction, targeting the criteria laid down. *ICE3004*, in the appendix, gives very clear guidance on the areas of your knowledge you need to demonstrate. You should make sure that both your synopsis and the eventual Technical Report show how you match the criteria.

Submission

The submission consists of a Technical Report, an Experience Report and a record of CPD. Remember that the overriding requirement for CPD is an average of five days per working year, so do not believe that the '30-day' specification for the conventional route necessarily applies to you. If you have many years' experience, it is likely that you no longer have a record of CPD from years ago; but at least show what you have done over the past few years, say six or seven as a minimum.

Technical Report

The requirement is for a report between 3000 and 10 000 words in length. In my experience as a Reviewer, the usual length seems to be around 5000 words. It must be an exposition of your major role in some aspect of civil engineering, showing how technical problems were resolved by the application of engineering principles and knowledge. It must show how your experience has compensated for your lack of formal education in the particular technical discipline you choose.

To do this successfully does, almost invariably, require you to do some swatting. Few engineers at this relatively advanced stage in their careers, are deeply involved in the day-to-day resolution of technical problems. Rather, you identify the problems and then delegate their detailed resolution to skilled technical colleagues.

So, if you do not do the detailed calculations yourself, then I believe it to be pretty certain that you do discuss the problem, help to identify possible solutions and have an input into the choice of the most appropriate one. To do this with credibility, you have to gain the trust of the colleagues who carry out the detailed analysis; they have to believe that you understand their difficulties and will not ask for anything stupid or unrealistic. So you *do* probably understand the technical principles to some extent. Now you really have to get to know the analytical details. You are not expected to be able to do the calculations, but you should develop sufficient understanding of the principles to be able to look at the solution and decide for yourself whether it 'looks right'. Your expert colleagues are the people to help you with this, so gain their cooperation.

You will also probably need to study relevant textbooks and, again, your colleagues will be able to refer you to the most suitable.

Nearly all successful candidates have found it necessary to read in depth and study around their subject area in order to be able to demonstrate an adequate understanding. It is rather like students doing their final academic examinations – during these, they are required to memorise and quote various formulae and equations which, under normal work situations, they would look up in an appropriate textbook.

You will be expected to be able to do something similar during your review. This process may seem a trifle artificial and unrealistic, but is it so very different from the academic situation you are being asked to replicate, where the average undergraduate swats for their final examinations?

The great danger when writing this report is that you inadvertently set out to prove that you are good at your job. A number of candidates have fallen into this trap and have been unsuccessful as a result. The Reviewers are probably convinced of your overall capability from your responsibilities at work. What they need you to demonstrate in this report is that you fully understand the engineering principles behind the solutions to some of your technical problems. You must ensure that your Technical Report demonstrates how you meet the criteria set out in the ICE guidance for the route.

Experience Report

This report requires you to deliver exactly the same attainments as a conventional candidate: 2000 words emphasising your burgeoning experience and contribution, clearly showing how you developed the attributes of a professional engineer. The latter part of this report will demonstrate how you used these attributes to drive projects to a successful conclusion. So it is a combination of the two reports written by conventional candidates, for which guidance is given in Chapters 12 and 13.

To be successful, you must use the greater part of the report to outline your current role and responsibilities, with just a brief sketch of the career which enabled you to reach that level of competence. It therefore seems sensible to write it in reverse chronological order, first describing what you are doing and then briefly outlining how you got there.

Summary of preparation

Throughout the preparation of these two components, you must keep in mind the requirements for a conventional candidate. In general, your Experience Report will tell the Institution what experience you have been exposed to; more importantly, it will tell them all the benefits you gained from that experience and how you now use the competencies you have developed. It is vital that you write this document with these targets in mind, otherwise you will follow a long list of people who have told the Institution only what they have done, but failed to tell them of the benefits they gained.

The Technical Report, on the other hand, is likely to be a more theoretical document, covering matters beyond your immediate day-to-day involvement. It is vital that you keep this distinction clear in your mind.

Much later on in the process, when your application is referred to the Engineering Council, you will be asked to write for them a brief explanation of why you believe you should be allowed to be a professional engineer. Surely it makes sense to take their description of an engineer (at the appropriate grade) and use this to describe yourself through your work? In this way you should cover all the aspects which the Engineering Council will be seeking. It is worth thinking about this, and perhaps producing a draft, at this early stage because, again, it will focus you on exactly what you are trying to prove.

Choosing your Lead Sponsor

You are asking this person to fulfil an onerous task, which will involve them in deciding for themselves whether you stand a reasonable chance of success and then reading and advising during the preparation of your documents and attending meetings with you. To fulfil these duties, they must not only have an up-to-date knowledge of the requirements of the appropriate grade of membership, but must also understand and appreciate the essential differences between the reports you are about to prepare and those prepared by conventional candidates. This latter aspect is vital; I have counselled several unsuccessful candidates who have received advice, given in the utmost good faith by established Reviewers of conventional candidates, but which has proven incorrect for this route.

If it is at all possible, the best sponsor is one who both knows the Institution's review criteria and is also an expert in the technical field you have chosen. The Institution can often help in identifying suitable persons, but it is up to you to approach them and gain their cooperation.

I strongly suggest that at an early stage, after you have selected a Lead Sponsor, you both arrange a meeting with the Institution's Regional Manager for your area, so that you can be sure that your proposed plan of action fully conforms with the system and that all parties are absolutely clear on what is to be demonstrated.

Review

There are three distinct parts to the TRR review:

❏ Academic Review,

❏ Professional Review,

❏ Written Test.

Academic Review

During this first review, your Reviewers will make sure that you have proved to them that you meet the criteria set out for the Technical Report and thus possess the required level of academic knowledge for the class of membership for which you are applying.

This review commences with a presentation by you, taking up to 30 minutes. This is the opportunity to extend the information you have already provided in your Technical Report. Do not simply repeat it. Half an hour is a relatively long time, and you must plan the content carefully to ensure that you display an in-depth understanding of the principles behind the report.

The presentation is followed by about an hour of questions, testing the range and depth of knowledge of the subject you chose for the

Technical Report. One of your Reviewers may well be an academic, and they will expect you to demonstrate the same understanding and knowledge of the technical principles as a conventional university graduate. They will not range over the full prospectus of a university, but will concentrate solely on the subject you have chosen.

There is then a short break, during which the two Reviewers will agree on whether or not you have convinced them. If not, the review will be terminated. If you have, then you will be recalled for the second interview.

Professional Review

Conditional upon success in the first part, the criteria for this interview are the same as for the conventional professional review. You may make another presentation, which will be very different from the one you gave earlier. This time you are reinforcing and expanding your case to become a professionally qualified engineer. You will then be questioned, just as a conventional candidate, until the Reviewers are convinced that you have all the attributes they are seeking for them to confirm that you are a professional engineer.

Written Test

You may be set a Written Test, at the Reviewers' discretion within ICE guidelines, dependent upon your experience. If you have more than 15 years' experience then you will not have to complete one. If, for example, it is clear from your role at work that you are capable of writing good reports under pressure and have submitted exemplary documentation, then the Reviewers may conclude that there is little reason to test this ability at review.

If you are required to do a Written Test, the specification is the same as for conventional candidates except that it will be carried out after your review and may be at a mutually convenient place remote from the review. Guidance on preparation for the Written Test is given in Chapter 19.

Summary

The burden of proof for the TRR is stringent. British professional engineering qualifications are frequently perceived within the European context as lacking in formal academic education, so there is constant pressure on the Engineering Council to prove otherwise. Those without any adequate academic qualifications can therefore expect to be examined in some considerable depth.

If the process is tackled logically and steadily, with advice and guidance at every stage, and the will to spend time to thoroughly understand the technical principles, then there is a very good chance of success for those who are professional civil engineers in all but qualification.

The steady publication of the names of successful candidates for formal election is strong evidence that the Technical Report Route does provide a real way forward for those who might otherwise have not been able to gain the same level of professional qualification.